T0130909

FERGUSON ADVERTISING

FERGUSON ADVERTISING

By John Farnworth

Herridge & Sons

ACKNOWLEDGEMENTS AND SOURCE OF MATERIAL

A considerable debt of gratitude is owed to AGCO and CNH Global N.V. (successor of the David Brown company) for allowing me to reproduce old advertisements.

Reading University Rural History Centre allowed me to have an extensive perusal of their archives from which I selected many advertisements to be copied. They then meticulously undertook this. Many of the old magazines have been bound into thick (often up to four inches), heavy volumes which made the copying process quite difficult in many cases. Many were bound before the days of photocopiers and the difficulty of copying from such large volumes could not have been foreseen. This has resulted in some copies inevitably showing a shadow effect along the bound margin. I should like to place on record my sincere thanks for their efforts because the Reading staff quite literally had to manhandle a great weight of bound volumes from archive shelves to distant copiers in the campus.

Much North American material was obtained by purchasing from individuals advertising on the Internet. These were obtained as individual pages which had been taken from newspapers and magazines. Some had been trimmed of page margins and in the process lost their dates.

David Lory, one of North America's leading Ferguson enthusiasts who has helped me source literature for my previous books again gave me valuable assistance in tracing old material – even at times prompting me about choice items currently for sale on the Internet! David also put me in contact with other collectors in USA who have made their contributions to the chapter start pages.

Mike Thorne sent me a couple of rare colour advertisements, and Rob Gilchrist explained to me the technical detail of the Ferguson "Suction Side Control" system of hydraulic control.

Thanks are due to all those who contributed to the chapter start pages which I hope has added a personal "collectors" dimension to the book. Sadly some of these are now deceased.

Considerable thanks are due to the publishers for the very professional manner in which they have designed this book, particularly taking into account that some of the original advertisements measured 13 by 20 inches!

Lastly, thanks are also due to my son Trystan and wife Moira who as always have helped with materials sourcing, preparation and organisation of the book. During the preparation of this book I have referred to:

Farnworth, J. *Massey Legacy. Vols. 1 and 2*

Farnworth, J. *Ferguson Implements and Accessories*

Fraser, C. *Harry Ferguson Inventor and Pioneer*

Gibbard, S. *The Ferguson Tractor Story*

Thorne, M. *Ferguson TE 20 Tractor in Detail*

Published in 2014 by
Herridge & Sons Ltd
Lower Forda, Shebbear
Devon EX21 5SY

ISBN 978-1-906133-62-7
Printed in China

CONTENTS

FOREWORD

BY JAMIE SHELDON

I think that even my grandfather Harry Ferguson would have been a little taken aback by the extent of Ferguson tractor and equipment advertising that John Farnworth has unearthed from more than half a century ago. And this collection only includes advertising in the English language – there would doubtless have been much more around the world in other languages and in particular in France.

This book is an excellent complement to John's previous book "The Advertising of Massey-Harris, Ferguson and Massey Ferguson." But whereas that was based on brochure material this new book is quite different in that it uses advertisements from magazines and newspapers. Advertising from these two sources is totally different in style. Those from newspapers and magazines can be very informative and often more casual in manner than brochure material. Quite a number were larger than brochures and this permitted the frequent use of "headline" effects in the overall presentation. Also, they were not as "glossy" as brochure advertisements but a small proportion were gloriously colourful.

John has quite logically divided the book into eight chapters to give specific coverage of each tractor model era from the Ferguson Brown through to the Ferguson 40 in North America. By starting each chapter with a brief coverage of the specific model's characteristics he shows how the Ferguson tractor evolved over the years. Most of the advertisements are highly informative about the tractor models, their implements and the uses to which they can be put. A significant number of the advertisements have been sourced from Australia and New Zealand and reflect the world wide sales that Ferguson System tractors and equipment received, and some odd uses to which they could be put such as rabbit control!

Although Harry Ferguson, his tractors and equipment have been widely written about, this new book adds a welcome and wholly new dimension to the overall subject.

Jamie Sheldon, 2014

LEADING THE WORLD

In the air, Britain's designs have brought new fame to British engineers and industrialists.

From just such engineers and industrialists comes the Ferguson System with its tractor and team of implements. This revolutionary System has blazed a new, spectacular trail throughout the world.

The System can make the good earth produce more than enough . . . and at prices which all can afford to pay. Thus we raise living standards by bringing down prices, and we destroy the seedbed of Moscow Communism. That is the course to which we are wholly dedicated.

HARRY FERGUSON LTD.

Ferguson tractors are manufactured for
Harry Ferguson Ltd., Coventry, by The Standard Motor Company Ltd.

3

Ferguson's advertising made many claims for its tractors and the Ferguson system, but an unusual one was to "destroy the seedbed of Moscow Communism"!

INTRODUCTION

In a previous book by myself – *The Advertising of Massey-Harris, Ferguson and Massey Ferguson* – many samples of Ferguson publicity brochures were reproduced as part of the long advertising history of Massey Ferguson and its predecessor companies. Other brochures were reproduced in *Fergusons. The Hunday Experience* which I prepared with John Moffitt. These brochures were essentially specially printed "hand-outs" (brochures, leaflets etc.) produced or commissioned by the companies, and given out to prospective purchasers by them, or by dealers, by special distribution (e.g. mail) or on request. This book has been well received so I decided to look more into the advertising of Ferguson. I was not disappointed – there is plenty to search out around the world.

An important advertising medium used extensively by Ferguson was journal and magazine advertising. Because of the very nature of newspaper and the associated print processes of the times, these advertisements are far less glossy than brochure material. They have much less colour, indeed many are black and white, and the text tends to have more of a journalistic style. As such, they are distinctly different to brochure material and have a character all of their own. They are very informative, often in a more casual manner than brochure material. Quite a number were larger than brochures and this permitted the frequent use of "headline" effects in the overall presentation.

Journal and magazine advertising in the Ferguson era had several advantages over the brochure advertising medium:

• It offered the most rapid way of presenting a new product to potential customers. Journals and magazines would typically be weekly or monthly. Mass audiences could be targeted and influenced at short notice and repeatedly.

• Many potential customers would "self expose" themselves to the advertisements in their every day life, i.e. through the routine purchases of journals or magazines. Potential customers therefore had to make little positive effort to acquire the product information.

• Many non-potential customers but nevertheless people who might influence a purchase e.g. farm workers, farmers' sons and farmers' wives, would incidentally read the advertising and then might enter into debate about the product with the potential purchasers, thereby prolonging the advertising effect.

• High impact advertising could be achieved by the headline approach, particularly with large format publications.

The advertising was in a more informal medium and therefore placed potential customers under less pressure to buy, but nevertheless the initial informing of the customer about a product was well achieved. By this process the potential customer is gently introduced to a product, hopefully causing him to subconsciously deliberate about its relevance to his circumstances prior to more formal approaches by salesmen.

• Journals and magazines tend to be read or browsed intermittently in the period before the next issue. This process repeatedly exposed potential customers to the product. In contrast brochures tend to be read once and then discarded or filed away.

• Many of the advertisements invited the potential customer to request further information (this is where the brochures would come in) or to have a demonstration of the equipment at which brochures would be given out. The journal or magazine advertisements were therefore often the first stage of an overall marketing ploy.

• Advertisements in more serious journals perhaps had an air of almost academic respectability because these types of journals often carried objective technical reports on advertised products as well.

The collecting of Ferguson brochure-type publicity material is now a well established part of the Ferguson enthusiast's hobby, together with instruction manuals and other types of literature. However collection of journal and magazine advertising material is still in its infancy. As far as I can make out it is a more developed activity in the USA than Europe. I hope that this book will inspire ever further diversification of Ferguson literature collecting.

I have organised the book into eight chapters. These are for the distinct tractor model periods.

The year of each advertisement is given where known; this may be derived from an actual date on a magazine or newspaper, or from a copyright date shown on the advertisement. Using this approach, the advertisements in each section have been presented in approximately chronological order. I have made a best guess to place advertisements when they were without date.

The advertisements cover the history of Ferguson tractors starting with the Ferguson Brown in 1936 through to the last Ferguson tractors – the Ferguson FE35s, TO35s and Ferguson 40 – in the late 1950s. However it is important to note that Harry Ferguson was experimenting with the prime principle of his "Ferguson System," that of implements being integrally attached to tractors effecting weight transfer from implement to tractor, from about 1917 onwards. Only two advertisements were traced for Ferguson's activities in the pre Ferguson Brown era. These are for the Ferguson single furrow plough used in the USA on Fordson tractors. One of his first demonstrations was in December 1917 in Northern Ireland featuring an Eros conversion Model T Ford car pulling a wheel-less plough. Ferguson tractors required Ferguson implements and this was the source of much of Ferguson's revenue rather than the tractors themselves. The Ferguson System may have started with just a plough but by the end of the Ferguson tractor era a vast range of Ferguson implements were being offered.

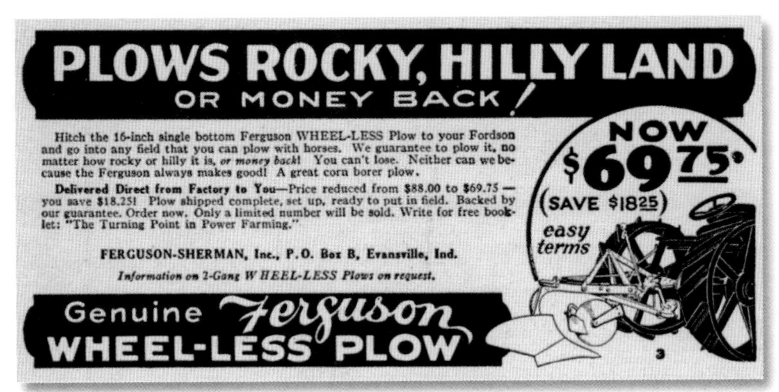

These are the only two advertisements found for the pre Ferguson Brown era. They are of North American origin and show a Ferguson single furrow plough being used on a Fordson tractor. This together with a two furrow plough was the first mass application of a Ferguson linkage system; many such ploughs were reputedly sold in the USA. However, as can be seen, development of the revolutionary weight transfer linkage at this stage was still not the three point type as we know it today, and was mechanical rather than hydraulically controlled.

ANTRIM FARMERS

PLEASE NOTE

A Tractor Ploughing

DEMONSTRATION

WILL BE GIVEN

In the Holm Field on the Land of Mr. WM. YOUNG,

FENAGHY, CULLYBACKEY,

ON THURSDAY NEXT,

Commencing at 11 o'clock in the Morning and 2 o'clock in the Afternoon.

At COLERAINE.

In conjunction with Messrs. STUART & CO., of BRIDGE STREET, COLERAINE, a similar DEMONSTRATION

WILL BE GIVEN

ON SATURDAY NEXT,

In the Field of Mr. JEWEL, DUNDOAN HOUSE,

COLERAINE,

Commencing at 10 o'clock in the Morning and 2 o'clock in the Afternoon.

As this Tractor and Plough are by far the most suitable for conditions in Ulster that have yet been exhibited, and as the Tractor will do all kinds of Agricultural work, and costs little more than half the usual price, we heartily invite all Farmers and all interested in Tractor development to these Demonstrations.

HARRY FERGUSON, LTD.,

NOTE ADDRESS MAY STREET, BELFAST.

ANTRIM FARMERS

Please Note!

A TRACTOR PLOUGHING DEMONSTRATION will be given in the Holm Field, on the Land of Mr. Wm. Young, Fenaghy, Cullybackey, on THURSDAY NEXT, commencing at 11 o'clock in the morning and 2 o'clock in the afternoon.

AT COLERAINE.

In conjunction with Messrs. Stuart & Co., of Bridge Street, Coleraine, a Similar DEMONSTRATION will be given on SATURDAY NEXT, in the field of Mr. Jewel, Dundoan House, Coleraine, commencing at 10 o'clock in the morning, and at 2 o'clock in the afternoon.

As this Tractor and Plough are by far the most suitable for conditions in Ulster that have yet been exhibited, and as the Tractor will do all kinds of Agricultural Work and costs little more than half the usual price, we heartily invite all farmers and all interested in Tractor Development to these Demonstrations.

HARRY FERGUSON, Ltd.

Note Address :- May St. BELFAST.

Newspaper advertising for a demonstration of the Ferguson plough in Northern Ireland in 1917.

Tractor Specifications

The following table gives a brief overview of the basic Ferguson tractor ranges covered in this book together with their production periods. They were essentially all made in either the UK or USA, but with a few TE variants in France. TE denoted "Tractor England" (likewise FE Ferguson England) and TO "Tractor Overseas".

The table shows how the first Ferguson tractor – the Ferguson Brown which was first produced in 1936 evolved over some 20 years to its final incarnation as the Ferguson 40 type tractor in 1956. An estimated 1,112,656 Ferguson system tractors were produced in the period with over a half a million of these being the UK built Ferguson TE tractors. When Massey Ferguson was created following the takeover of Ferguson by Massey-Harris, production of the FE 35, TO 35 and Ferguson 40 tractors carried on but with MF livery and badging, so the actual number of "Ferguson" style tractors was significantly higher than the 1.1 million indicated. It is also noted that the Ferguson 40 tractor was first produced as the Massey-Harris 50 tractor starting some six months earlier in response to M-H dealers craving for a Ferguson system tractor. They differed only in their styling and livery and 15,707 of them were produced prior to the advent of Massey Ferguson.

	Ferguson Brown	Ferguson TE 20	Ferguson FE 35	Ford Ferguson	Ferguson TO 20	Ferguson TO 30	Ferguson TO 35	Ferguson F 40
Production period	1936-1939	1946-1956	1956-1957	1939-1947	1948-1951	1951-1954	1954-1960	1956-1957
No. produced	About 1,354	517,651	74,655	306,221	About 60,000	About 80,000	About 63,678	9,097
Manufacturing location	UK	UK	UK	USA	USA	USA	USA	USA
Hp	20 bhp	23.9-28.2 hp	29-37 bhp	17.5-23.8 max belt hp	26.5 max belt hp	"2-3 plough capacity"	"3 plough capacity"	"3 plough capacity"
Engines – all 4 cylinders	Hercules, later Coventry Climax	Standard 80 and 85mm, Continental Z120 or Standard 20C diesel (Perkins three-cylinder P3TA diesel for Yugoslavia only)	Standard 23C diesel or Standard 87mm	Ford	Continental Z120	Continental Z130	Continental Z134 or Standard 23C diesel	Continental Z134
Fuel	Petrol Petrol/TVO Diesel Lamp oil	Petrol Petrol/TVO Diesel Lamp oil	Petrol Petrol/TVO	Petrol	Petrol	Petrol	Petrol Diesel	Petrol
Gear ranges	1	1 (Ferguson epicyclic reduction gearbox available)	2	1	1	1	2	2
Total gears	3F + 1R	4F + 1R	6F + 2R	3F + 1R	4F + 1R	4F + 1R	6F + 2R	6F + 2R
Max speed forward	4.9 mph	13.25 mph	14 mph	6.62mph @ 1400rpm	9.84mph @ 1500rpm	9.84 mph @ 1500rpm	10.95mph @ 1500rpm	14.6-16.5mph depending on tyres
Weight	About 1848lb	2376-2702lb	2982-3158lb	2140lb	2500lb	2570lb	2900lb	3100-3280 lb
Wheel/chassis configuration	4 wheel	4 wheel	4 wheel	4 wheel	4 wheel	4 wheel	4 wheel	4 wheel, V twin and single front. Standard and Hi-arch
Ignition	Magneto	6V and 12V	12V	Magneto and 6V	6V	6V	6V and later 12V	12V

THE FERGUSON BROWN

The Ferguson Brown or Model A tractor was Harry Ferguson's first production Ferguson System tractor. His very first prototype tractor was the Ferguson Black tractor which is preserved in the Science Museum in London. It was a revolution in tractor design in that tractor and implement came together as an integrated concept for the first time as the "Ferguson System". A major part of the Ferguson System was that the implement was attached to the tractor by a hydraulically actuated three point linkage mechanism which lowered and raised the implement in and out of work, automatically controlled its working depth and effected weight transfer from the implement's operation to the rear of the tractor to improve traction. This latter feature enabled use of a light weight tractor and implement to do the same work as previously by much heavier tractors and implements.

The Ferguson Brown tractor was made for Ferguson by the David Brown company but the agreement was not to last. Ferguson then went to the USA and arranged production of the Ford Ferguson tractor by Henry Ford. Although sales of the Ferguson Brown were limited it did prove the whole principle of a Ferguson System tractor and international sales were achieved. Besides the standard agricultural version a very limited number of orchard and industrial models were also made. Perhaps the biggest weakness of this first Ferguson tractor was that the tractor actually had to be in motion for the hydraulic lift to work. A major difference between the prototype Ferguson Black tractor and the Ferguson Browns was that they had the hydraulic control valve on the suction side of the pump – "suction side control" instead of on the output side. An additional feature of this system was a major safety factor – if the implement hit an obstacle then automatic overload release prevented the tractor from rearing up. This feature came about almost accidentally as a consequence of the re-design of the pump to suction side control. Prior to this there had been problems with the Ferguson Black tractor suffering air locks in the hydraulics and overheating of the oil.

The dawn of a new security and prosperity for Britain

THE series of inventions incorporated in the Ferguson Hydraulic Agricultural Machinery are opening up vast possibilities for British agriculture and industry. They are revolutionizing cultivation and will make farming highly profitable, because they enable food stuffs to be produced at greatly decreased cost.

The horse is an old friend, but as a source of power he is slow and expensive. Work previously done by horses can now be carried out with much greater speed and efficiency, whether the farm be small or large, and the easy, interesting methods of operation will attract youth and capital to the land.

It is vital that home productions should be increased to a level which will make us independent of imported food supplies during a national emergency. We have not sufficient land to feed our population plus the millions of horses required to make this country self-supporting.

The Ferguson Machinery will produce at half the cost of any other method and is the solution of our agricultural problem. The farmer can now make large profits, and new opportunities of healthy and prosperous careers are opened up for thousands of our people.

With the advent of the Ferguson Machinery the danger of a shortage of food supplies in a national emergency can no longer be justified. Britain can now produce all she needs; safeguard herself in war, and provide permanent work for her unemployed.

The Ferguson inventions have been deliberately planned and perfected to achieve these ends.

AT SMITHFIELD SHOW — STAND 18

Possibly one of the earliest advertisements for the Ferguson Brown and Ferguson machinery. It would certainly have been Ferguson's first appearance at the pre-Christmas Smithfield show. November 1936.

This is the first advertisement found for the Ferguson Brown era and dated February 9th 1937. Note the well-dressed driver – Harry Ferguson always insisted on his demonstrators presenting themselves and the machines immaculately. He started as he intended to continue. Strange that this early advertisement should be for splitting the potato ridges – perhaps anticipation of the imminent potato planting season?

A joint appeal to farmers and dealers to come forward and buy. Note the early tractor covers marked "Ferguson Hydraulic Farm Machinery". March 1937.

Immaculate ridges from a Ferguson Brown tolling the death knell for horse power. March 1937.

This seems to be the first advertisement to show a Ferguson Brown with dirty wheels and a working farmer driving. The destruction proof nature offered by the combination of spring loaded tines and the Ferguson hydraulic control system is claimed in this field full of boulders. March 1937.

1062 THE FARMER AND STOCK-BREEDER. April 27, 1937

LABOUR PROBLEMS SOLVED!!

Greatly increased profits enable higher wages to be paid. You will have the choice of the best tractor operators, because the FERGUSON is so interesting, so modern, and so easy to work.

Ploughing with the Ferguson will increase the productivity of your land.

Farmers everywhere state that the Ferguson patented unbreakable spring tine Cultivator does the best work they have ever seen.

The Ferguson can beat horses even in setting up and splitting ridges, it carries out a great deal more work at a much lower cost—let it make handsome profits for you.

An amazing achievement, the Ferguson implement cannot be damaged by striking obstructions.

Small fields and awkward corners are negotiated with ease owing to the patented unit construction.

Ferguson Farm Machinery with hydraulic implement control requires smaller headland than horses.

You can operate the FERGUSON where all other tractors fail—on hills—on hill-sides—when the land is wet—and among obstructions. Ideal for market gardens, fruit and vegetable farms, orchards, beet cultivation, hop-yards, etc.

Apply for full particulars to:

HARRY FERGUSON, Ltd.,

Telephone: 2685. **HUDDERSFIELD.** Telegrams: "FARMING."

Quality work, precise manoeuvrability, breakage resistance and labour saving. Harry Ferguson lays his claim to mechanisation superiority. April 1937.

May 1, 1937. THE IMPLEMENT AND MACHINERY REVIEW. 27

This spells "MONOPOLY" for the "DISTRIBUTOR"

EASE & SIMPLICITY OF OPERATION

AUTOMATIC DEPTH CONTROL

UNIQUE THREE POINT HITCH

Are **YOU** selling

FERGUSON FARM MACHINERY?

With Hydraulically operated and controlled implements.

HYDRAULICS · HARNESSED FOR THE FARMER

20 H.P

HIGH POWER WEIGHT RATIO

HANDSOME PROFITS FOR THE TRADE

For particulars apply to HARRY FERGUSON L^TD HUDDERSFIELD

Phone - 2685 Grams-FARMING

More dealers needed. Critical marketing design features emphasised. This is possibly the first advertisement with the tractor on rubber tyres. May 1937.

276 THE IMPLEMENT AND MACHINERY REVIEW. July 1, 1937.

TO AGRICULTURAL ENGINEERS

The above has appeared in the National Agricultural Press. It is worth your investigation. The Ferguson means profitable business for you too.

The Ferguson Brown goes to the Royal Show for possibly the first time. July 1937.

1738 *THE FARMER AND STOCK-BREEDER.* July 13, 1937.

FERGUSON MACHINERY

—the equipment that will revolutionise agriculture—was the centre of attraction at THE ROYAL SHOW.

Have you placed your ORDER?

1 Ploughing with the Ferguson will increase the productivity of your land; use this equipment and increase your profits greatly.

2 Farmers everywhere state that the Ferguson patented unbreakable spring tine Cultivator does the best work they have ever seen.

3 The Ferguson can beat horses even in setting up and splitting ridges; it carries out a great deal more work at half the cost.

4 An amazing achievement, the Ferguson implement cannot be damaged by striking obstructions.

5 In small fields and awkward corners the horse is completely eclipsed for ease of handling by the Ferguson unit equipment.

6 For Row Crop Cultivation of all kinds the Ferguson is supreme. With one lad to operate, it will do as much work in one day as four men and eight horses.

You can operate the Ferguson where all other Tractors fail—on Hills—on Hillsides—on Wet Land—Amongst Obstructions. Ideal for any Size of Farm from 20 Acres upwards, also for Fruit Farms and Market Gardens

Telephone: 2685

HARRY FERGUSON, LTD.
HUDDERSFIELD

Telegrams: "Farming"

Selling the versatility of the Ferguson Brown. July 1937.

Could dealers resist the lure of a product giving them a monopoly on what was to become a world beating concept? August 1937.

AUGUST 1, 1937. THE IMPLEMENT AND MACHINERY REVIEW. 373

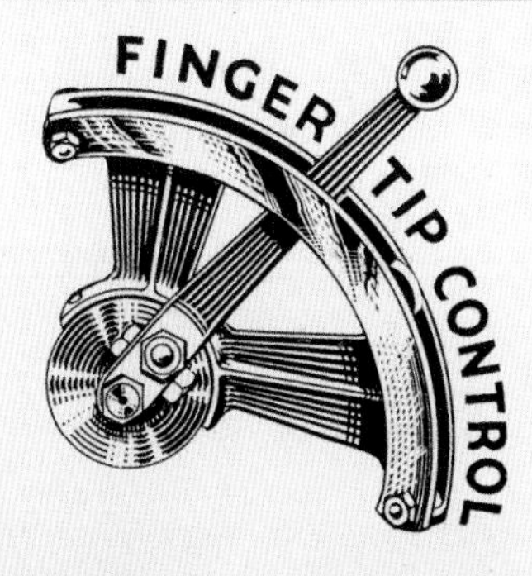

FEATURES THAT SPELL MONOPOLY FOR THE DISTRIBUTOR . . .

ARE YOU SELLING FERGUSON FARM MACHINERY

The Ferguson is the answer to the farmer's labour problems. By giving him reliable and easily controlled equipment, it turns farming from heavy labour into a pleasurable occupation.

Besides being exceedingly simple to control, it is such a fine engineering job it appeals strongly to the operator.

The design of the Ferguson is carried out so that skilled labour is unnecessary. Excellent work may be carried out by semi-skilled or unskilled men.

Space and time are saved on headlands, and exceedingly light weight makes running expenses low.

Less work and greatly increased profits are provided for every Ferguson user.

AMAZINGLY SIMPLE TO OPERATE

HARRY FERGUSON LTD. HUDDERSFIELD

Phone:-HUDDERSFIELD 2685 *Grams:-FARMING HUDDERSFIELD*

470 THE IMPLEMENT AND MACHINERY REVIEW. SEPT. 1, 1937.

HYDRAULIC IMPLEMENT CONTROL

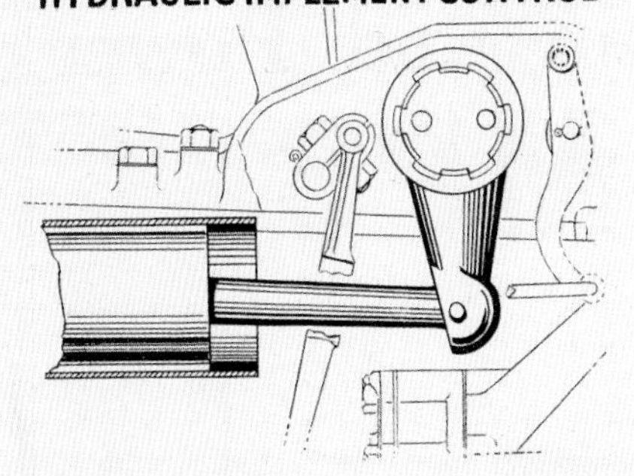

FEATURES THAT SPELL MONOPOLY FOR THE DISTRIBUTOR . . .

ARE YOU SELLING FERGUSON FARM MACHINERY

Wonderfully simple—hydraulic implement control—all that is required to drop or lift the Ferguson Patented Implements is the movement of a finger-tip lever. And that is not all —implements remain automatically at the depth they are set, even when the tractor is pitching over uneven ground.

Again, there is a great saving in both time and space on headlands, as the implement is carried direct at the rear of the tractor.

Then there is the weight question. The Ferguson construction enables all useless weight to be dispensed with. This saves fuel and upkeep costs, and eliminates land compression.

The Ferguson is worth investigation.

HARRY FERGUSON LTD. HUDDERSFIELD

Phone :-HUDDERSFIELD 2685 Grams:-FARMING HUDDERSFIELD

Another appeal to dealers. Look at the size of the field and quality of the work. September 1937.

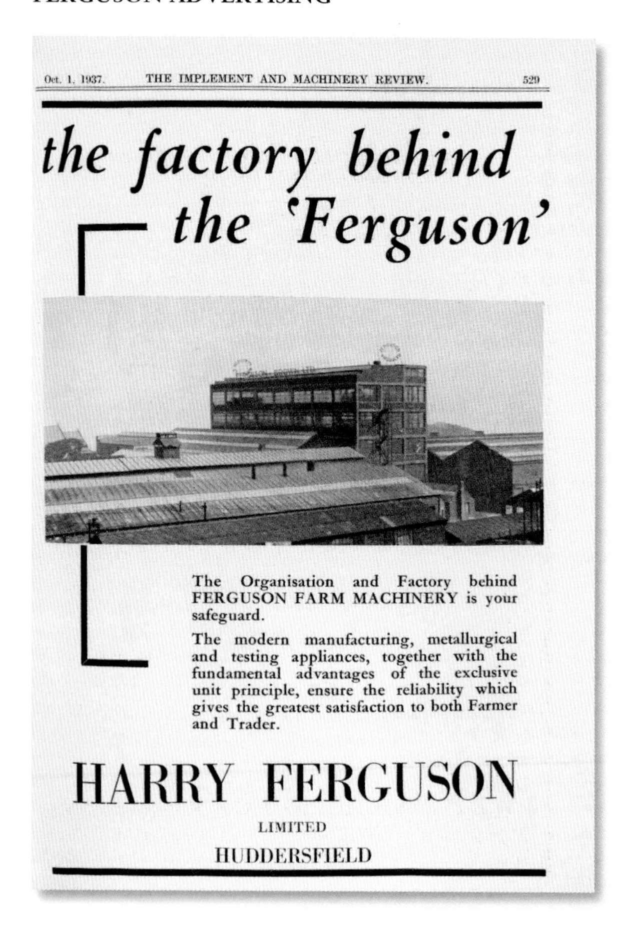

Perhaps the first advertisement to show the connection with David Brown emphasises the quality of engineering going into the Ferguson Farm Machinery. The "Ferguson Brown" tractor was made for Ferguson by David Brown. October 1937.

JAN. 1, 1938. THE IMPLEMENT AND MACHINERY REVIEW. 853

FERGUSON FARM MACHINERY

The "Ferguson" equipped with pneumatic tyres, mudguards, and with two-furrow plough attached.

If you were unable to visit Smithfield Show and consequently missed the opportunity of seeing this scientific machinery, which is making farming into a profitable Industry, write for full particulars from the makers.

Export enquiries invited.

FERGUSON-BROWN L^{TD}

HUDDERSFIELD

Possibly the first advertisement of the Ferguson Brown with mudguards. Note Harry Ferguson going for the export market. January 1938.

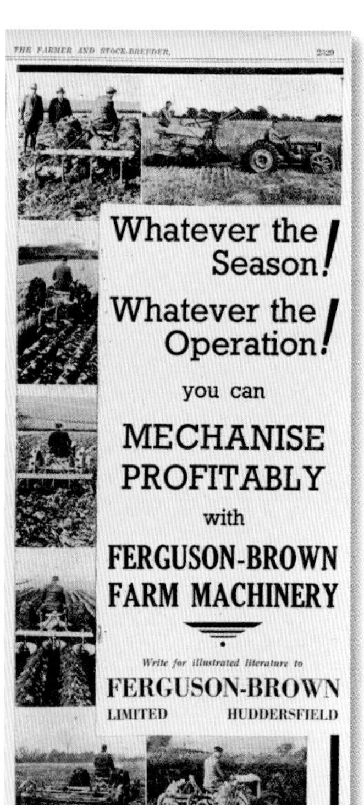

This might be the first advertisement showing the Ferguson Brown using non Ferguson trailed implements. It may also be the first to show the use of the new company name – Ferguson-Brown. October 1937.

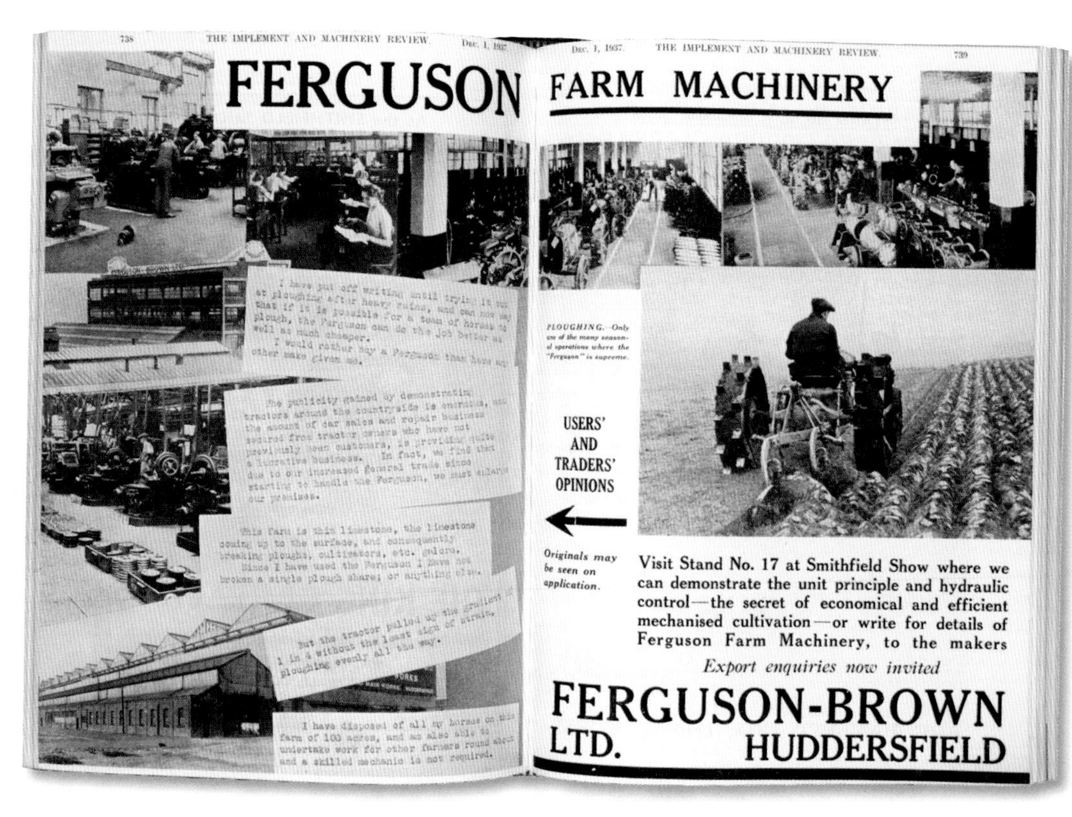

This appears to be the first double page advertisement. Farmers' experiences and praise starting to be used to promote the product against the background of the efficient factory production complex and a shot of quality ploughing. December 1937.

Another attack on the poor old horses! January 1938.

100 THE FARMER AND STOCK-BREEDER.

Growing profits with
FERGUSON
Farm Machinery

Successful farming depends on taking immediate advantage of the frequently brief opportunities afforded by the weather. This is where the "Ferguson" scores. Its performance is equivalent to about 200 horse/hours per week, while the scientific distribution of its weight and the fact that the implements float behind the tractor instead of being hauled, enable it to operate when work is impossible by any other method. Start your season profitably by cultivating and harrowing with FERGUSON FARM MACHINERY.

FERGUSON - BROWN
LTD HUDDERSFIELD

Still looking for dealers and farmers to buy this exclusive product. February 1938.

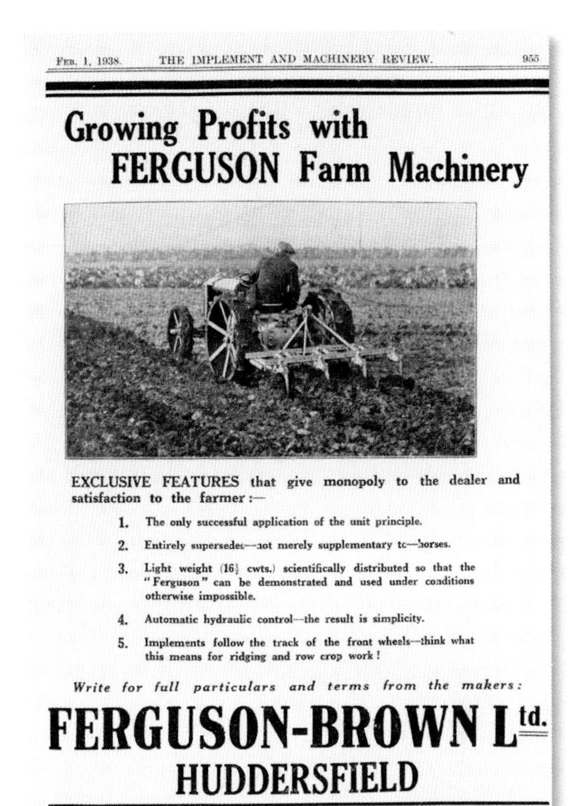

FEB. 1, 1938. THE IMPLEMENT AND MACHINERY REVIEW. 955
Growing Profits with
FERGUSON Farm Machinery
EXCLUSIVE FEATURES that give monopoly to the dealer and satisfaction to the farmer :—
1. The only successful application of the unit principle.
2. Entirely supersedes—not merely supplementary to—horses.
3. Light weight (16½ cwts.) scientifically distributed so that the "Ferguson" can be demonstrated and used under conditions otherwise impossible.
4. Automatic hydraulic control—the result is simplicity.
5. Implements follow the track of the front wheels—think what this means for ridging and row crop work!
Write for full particulars and terms from the makers:
FERGUSON-BROWN Ltd.
HUDDERSFIELD

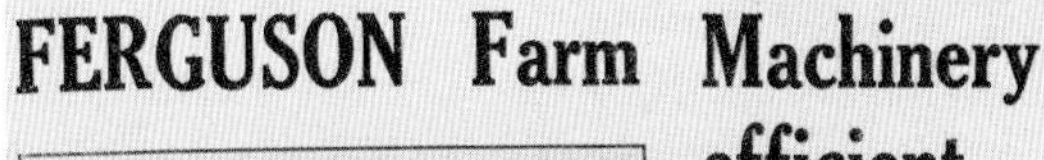

FERGUSON Farm Machinery

efficient
and
profitable

HORSES
picturesque
but
out of date

Outstanding features of Ferguson Farm Machinery :—

- Automatic depth control.
- Implements can be changed in one minute.
- Patent hitch by which the implements track with the FRONT wheels.
- Does not pack the land.
- Does not merely supplement, but entirely supersedes horses.

FERGUSON-BROWN
LIMITED HUDDERSFIELD

The horses would struggle to produce a finish to the ploughing like the Ferguson Brown. The horse's main claim to superiority – not packing the land – under attack. March 1938.

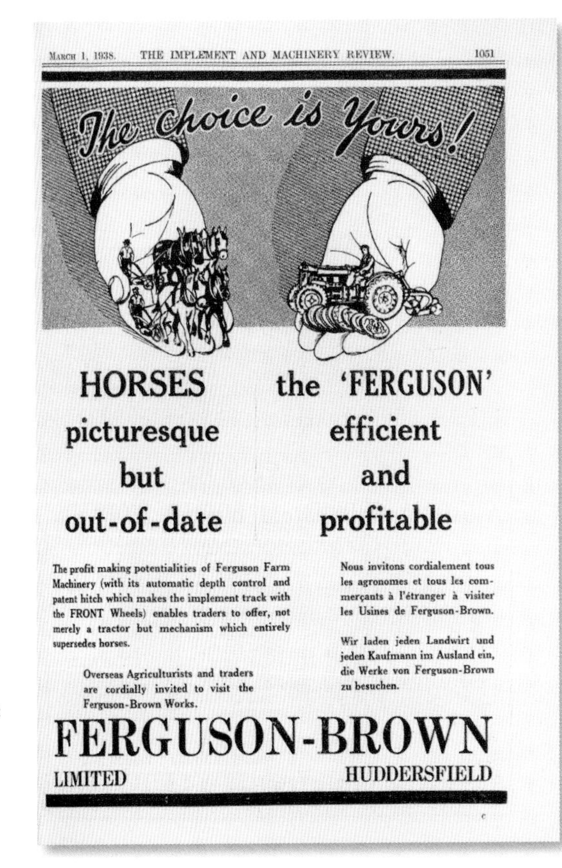

MARCH 1, 1938. THE IMPLEMENT AND MACHINERY REVIEW. 1051
The Choice is Yours!
HORSES
picturesque
but
out-of-date
the 'FERGUSON'
efficient
and
profitable
The profit making potentialities of Ferguson Farm Machinery (with its automatic depth control and patent hitch which makes the implement track with the FRONT Wheels) enables traders to offer, not merely a tractor but mechanism which entirely supersedes horses.
Overseas Agriculturists and traders are cordially invited to visit the Ferguson-Brown Works.
Nous invitons cordialement tous les agronomes et tous les commerçants à l'étranger à visiter les Usines de Ferguson-Brown.
Wir laden jeden Landwirt und jeden Kaufmann im Ausland ein, die Werke von Ferguson-Brown zu besuchen.
FERGUSON-BROWN
LIMITED HUDDERSFIELD

Multi-lingual appeal to give up horses, and for experts and traders to assess the tractors at the factory. March 1938.

APRIL 1, 1938. THE IMPLEMENT AND MACHINERY REVIEW. 1159

ROW CROP WORK
Ridging and Splitting Ridges

with FERGUSON Farm Machinery

Ferguson Farm Machinery, on account of its exclusive features whereby the implements follow the track of the FRONT WHEELS, entirely supersedes horses for all row crop work, including Ridging, Cultivating and Splitting Ridges.
The seasonal work now in progress offers Traders unique opportunities for demonstrating the profit-making possibilities of this efficient and robust mechanism. Export enquiries are now invited.

FERGUSON-BROWN
LIMITED HUDDERSFIELD

A seasonal plea to adopt Ferguson Farm Machinery for row crop work. April 1938.

Ridge splitting was always one of the most difficult row crop jobs, the driver looks tense even on the Ferguson Brown! May 1938.

Produce luxuriant potato crops with a Ferguson Brown. June 1938.

Haul home the harvest with a Ferguson Brown, but note that 30 percent of cultivation tasks are still deemed to need steel wheels. August 1938.

Exports taking off. A significant proportion of the Ferguson Browns produced were exported. September 1938.

A view of the plough that the world came to know so well. November 1938.

A clean and relaxed driver and immaculate workmanship. Note the reduced size of fenders compared to those shown in the January advertisement. October 1938.

See us at the Royal Smithfield Show for a full explanation of our superiority. Also possibly the first mention of the Ferguson training facilities. December 1938.

A New Year appeal, from January 1st 1939, to adopt the Ferguson System. It may be the first advert showing the narrow row crop rear wheels. Below it is a repeat from February 1939, but note the company name change to David Brown Tractors.

Possibly the last advertisement for the Ferguson Brown tractor still inviting export enquiries. April 1939.

APRIL 1, 1939. THE IMPLEMENT AND MACHINERY REVIEW. 1159

If your customers have cultivating jobs to do—and do well—give them a demonstration of the UNIT PRINCIPLE on their farm, for the Ferguson, with its many exclusive features,

COVERS THE WHOLE FIELD of Cultivation

FLOATING IMPLEMENTS, HYDRAULIC CONTROL, PATENT LINKAGE and LIGHT WEIGHT are points of superiority with enormous possibilities in your district. Write for particulars now !

★ EXPORT ENQUIRIES INVITED ★

DAVID BROWN TRACTORS
LIMITED HUDDERSFIELD
FORMERLY FERGUSON-BROWN LIMITED

THE FORD FERGUSON

Sales of the Ford tractor with Ferguson System – the Ford Ferguson – were truly massive compared with its predecessor Ferguson Brown. In the Ford Ferguson tractor the Ferguson System tractor can truly be said to have "taken off". It was a somewhat heavier and slightly more powerful tractor and with the important improvement of the hydraulics operating when the tractor was stationary but with the clutch engaged. This enabled implements to be lifted whilst the tractor was stationary. The tractor was produced largely over the course of World War II and reportedly at least 10,000 of them came across the Atlantic to the UK and Europe under the wartime lend-lease agreement with the USA. In the UK they were never on open sale but rather distributed by government to farmers. A much wider range of implements were available for the Ford Fergusons – particularly in North America. Orchard and industrial models were also made in limited numbers as well as "tugs" for use in military situations. Wartime economy (utility) models were produced with magneto ignition and steel wheels. Production of the Ford Ferguson tractor benefitted from Ford's mass automobile production experiences and this included the fitting of a self starter. During the war Harry Ferguson made a limited range of implements for the Ford Fergusons at his factory in Moira near Belfast. In 1947 Ford broke with Ferguson and the production of Ford Fergusons discontinued. This left Harry Ferguson with no manufacturer for his tractors.

Due to wartime constraints and government control of distribution there does not appear to have been any advertising of the Ford Ferguson in the UK.

1941 • FARM JOURNAL and FARMER'S WIFE 5

THE FUTURE STARTS TOMORROW

Each tomorrow confronts us with new conditions which affect the way we organize our life and work.

How we organize tells whether or not we shall be master of those conditions.

NO ONE KNOWS exactly what the morrow will bring.

We're probably going to have to feed the world, but at what price?

Who can say the relative levels at which the farmer is going to have to buy what he needs and sell what he raises?

What temptation of apparent high wages and easy hours is the city going to offer to farm youth, to your boy and girl?

What new lullaby is going to lull the farmer into forgetfulness of the old American principles of self-determination and the satisfactions and rewards of personal initiative?

How much more in farm produce is going to be taken by industry, and how soon?

To these and other world forces there is no A B C answer. But one thing is as plain as can be:

The farmer who drives his costs down with modern equipment is the man who is going to make the money, no matter what happens.

Also, the farmer who makes farming attractive to his son is the man who is going to keep him on the farm, where his roots are, and where you need him and America needs him.

You, yourself, deserve to have the day's work really done before the next day begins. You deserve leisure, and the pep and the clear conscience to enjoy it, and some cash money to spend.

It's been up to someone to provide the means.

We honestly believe we have done so. Therefore we make no apologies for a strictly commercial recommendation.

You are invited to see how we have designed and built the Ford Tractor with Ferguson System to lick the future for you and your son.

Why do we dare make such a strong statement? Because this System is different. It is different in principle. It applies power a new way. It saves time and drudgery and it makes profits. It does make farming attractive to youth for exactly those reasons.

The kind of future you want can start tomorrow on the comfortable seat of this modern machine.

There is a Ford Tractor dealer near you. He will gladly prove what we have said—on your own farm.

© Ford Motor Co.

The Ford Tractor with Ferguson System is sold nationally by the Ferguson-Sherman Manufacturing Corporation, Dearborn, Michigan, and distributed through dealers in every part of the country.

A youth fascinated by the prospect of advancing farm mechanization with the advent of the Ford Ferguson tractor. 1941.

49

TWO MEN WITH A SINGLE PURPOSE

I have long held the conviction that something should be done about farming. In too many cases, farming has not only ceased to be profitable; it has also ceased to be interesting.

The land itself has not collapsed either in expanse or productivity. But means have not been at hand whereby the family unit, on which the well-being of the land must depend, could produce at a profit and at the same time have the leisure to enjoy the fruits of their labor.

When Mr. Ferguson approached the problem of mechanization in terms of fundamental principles, and solved it in terms of the average farm family, our purposes became as one.

We both believe farming can and must be made profitable.

That is why we introduced the new Ford tractor just a few brief months ago.

Henry Ford

"Now 37,283 Farmer-Owners know we have the Answer"

Our purpose in developing the entirely new system of mechanization which is embodied in the new Ford tractor, was to make available at low cost the means to accomplish the four imperatives to a prosperous agriculture.

These four principles are set forth at the right.

That they have been brought within the reach of every American Farmer so cheaply, and so quickly after the perfection of the system, is a tribute to the manufacturing genius of Mr. Ford, and to his great belief in the land.

Harry Ferguson

1 To cut the actual cost of crop production on the family farm. Not just to do certain special things in a spectacular way, but to do all kinds of farming on all kinds of farms more cheaply than it had ever been done before with *anything.*

37,283 farmer-owners now know we have the answer

2 To make farming attractive to youth, and easier for every member of the family. The drift from the land will stop when our *young men* have farming equipment which is both *capable* and interesting, and when they can confidently look forward to a profit.

37,283 farmer-owners now know we have the answer

3 To assist all other industries through a prosperous agriculture. As the farmer profits, the country prospers; as the country prospers, the farmer profits. It is an endless cycle, but it cannot begin until the individual farmer can produce for less money than he can sell for.

37,283 farmer-owners now know we have the answer

4 To lay the foundation of a greater National security. From the land comes everything that supports the life of all our 130,000,000 people, and half of these depend directly upon the farm for livelihood. If the farmer himself can produce at a profit without raising the price to the consumer, nothing can destroy the security of this country.

37,283 farmer-owners now know we have the answer

The Ford tractor with Ferguson system is sold nationally by the Ferguson-Sherman Manufacturing Corporation, Dearborn, Michigan, and distributed through dealers in every part of the country.

SUCCESSFUL FARMING, FEBRUARY, 1941 — SEE . . . . PAGE 111

Two great men came together to produce the first truly mass produced Ferguson System tractor, but strictly speaking it was a Ford tractor with Ferguson System. Ford produced and Ferguson-Sherman marketed the tractor. Many implements were also made through the Sherman link. February 1941.

1941 • FARM JOURNAL and FARMER'S WIFE 5

THEY CAN TEACH, TOO!

We grown-ups have no exclusive claim on wisdom.

Youth has a way of looking at life and seeing through it with crystal clarity.

There are whole volumes of farm economics to be learned from the simple fact that farm youth is definitely machine-minded.

YOUNG MEN AND WOMEN on the farm are hearing their elders talk a lot about the youth problem: about the pity of the drift from the land; about "selling" youth the future of farming.

You know the kind of reply youth is making to all that.

Educational and inspirational work of superlative value is being offered by our agricultural colleges, their extension divisions, and "short-courses"; by Smith-Hughes schools; by the Future Farmers of America, the 4-H Clubs, the National Farm Youth Foundation, and others.

But, when seven out of eight farm youths go home from school or class, it is to a farm which is operated with a four-legged power plant so inefficient that it consumes the produce of one acre out of five, so slow that the hours needed in the fields seem to drag out to eternity.

No wonder youth wonders about a future filled with the long hours, the drudgery, the endless chores that never can be avoided with horse-farming. Not to mention the drain on income.

Youth sees all the world mechanized but farming, and he wants to be a part of that kind of world.

Machines do better work, cheaper, for men in other occupations. . . .

Machines give other men profit, and the leisure to enjoy it.

Listen closely to the heartbeats of your own boy, and if you are *very* understanding, you will learn that his real urge is to do the thing he knows most about, farming—provided he can do it the modern way, with the machine.

He can, now.

There is available for the first time a machine which is *actually modern* because, by discarding all the old ideas about pulling and controlling implements in the soil, it does away with all need for the horse.

The Ford Tractor with Ferguson System is *not* just another good motor to be hitched in front of the same old implements. It is *not* a tractor made to replace horses only under ideal and limited conditions. It is a new SYSTEM of land cultivation.

In the Ford Tractor with Ferguson System power is applied through a new principle which makes tractor and implement a single operating unit. The System is so efficient that this light tractor, using little fuel, gets *all your work* done with ease, **at a cost that enables you to make sound profits, and that provides the leisure to spend them.**

This is the machine of which youth is saying, "Let Dad look into the costs, but I know a machine when I see it."

The Ford Tractor with Ferguson System is sold nationally by the Ferguson-Sherman Manufacturing Corporation, Dearborn, Michigan, and distributed through dealers in every part of the country.

Encouraging youth to become involved with farm mechanisation and replacement of horses. And, emphasising the new system approach to mechanisation. March 1941.

A simple but lengthy message. Get rid of the horses, get your son driving, enjoy some leisure time in the garden – just ask for a demonstration. September 1941.

COUNTRY GENTLEMAN 1

NOT HOW BIG . . . BUT HOW WELL

If we could find out who it is that owns this farm . . .
this is the message we would write him, personally:

"Dear Sir: You have some acres, some buildings and a family, perhaps a boy and a girl.

"You've been having some good years and some bad ones, and you're wondering how you ought to be planning things, not only for more year-by-year security, but so you might be laying a few dollars aside.

"It wouldn't be surprising if you had been tempted by the idea of another farm, more acres, to solve the problem.

"What you've got now looks pretty good to us. Over there by the barn is a nice piece of pasture. That means cows, but it looks like horses, too.

"As man to man, why on earth horses? Are they efficient? Are they fast? Do you really like to give up five acres of pasture and crops to feed each horse you work?

"Can it be that you are one of the farmers who say . . . 'Sure, I know horses are slow and use up a lot of my land, but I haven't seen any tractor yet that would do all the work my horses do, and work cheaply enough to let me get rid of my horses'.

"Then, sir, this is our personal promise to you that unless you have farmed the land you've got with our kind of machine, you haven't even touched the possibilities of that land.

"We have made a new kind of machine. It isn't like any tractor you ever tried before. It is so light it will not pack the soil. It's not a big machine to look at, or to buy gas for, yet it will break heavy sod with two fourteen-inch plows under any normal conditions.

"It will do more kinds of farming, faster and cheaper, and better than you ever dreamed possible, with machines or horses, or both.

"With what you know about farming, and a machine like that to do it with, you can make those acres you've got, pay as they have never paid before.

"That boy of yours, even if he is only a youngster, can run this machine, and stack his results right up against yours . . . it's that simple and easy to operate.

"And by the way, we see no garden-patch in this picture. Are we far off when we guess that you told your good wife you just didn't have time to bother with one? Time is one big thing this machine saves you. Farming the way we are suggesting will give you time to raise the garden that will save you sending many a dollar off to market.

"And that's all we have room to say here. And since we don't know your name, nor where you live, the only way we can show you this machine is to ask you to go to the nearest dealer who sells the Ford Tractor with Ferguson System, and ask him to bring one out to your place, so you can see with your own eyes what we mean."

The Ferguson Mower cuts large or small fields and odd corners. Easy to operate. Finger tip hydraulic control of Ferguson Implements takes the drudgery out of farm work.

© Ford Motor Co.

The Ford Tractor with Ferguson System is sold nationally by the Ferguson-Sherman Manufacturing Corporation, Dearborn, Michigan, and distributed through dealers in every part of the country.

COUNTRY GENTLEMAN 33

14,000,000 BELOVED CULPRITS

THERE IS SOMETHING about man's love for a horse, for his horse, that weighs against his ever being quite cold-blooded in judgment about the horse as a machine for doing farm work.

Farming is our biggest industry. More than one third of all property used to produce, in America, is devoted to agriculture. Yet all other industries long since discovered that the horse is less efficient than the machine. Only the farm holds on.

We think there are two reasons:

FIRST. The farmer thinks it is easy and inexpensive to feed his horses, which is to say, that horse-fuel is cheap.

That deserves a look. A horse consumes, on the average, the produce of a little better than 5 acres. Is capable of working 22 acres. The terrible fact is that *one acre out of every four* worked by the horse belongs to the horse, not to the farmer.

Worse than that. A farmer averages one full month of 30 ten-hour days every year just taking care of his horse. That is a whole April stolen from productive work.

No farmer who reads this page would willingly give up a quarter of his farm, strike April out of his calendar, and count on success. Yet that is what every man who works his farm with horses does every year.

Seven out of eight farms still operate with horses. Six farm families out of eight have an income of less than $1,500 per year. It is a deadly parallel.

SECOND. Until now, the farmer couldn't do without horses, even if he wanted to.

Inefficient, slow though he is, the horse *is* versatile. The real reason there are still over 14 million horses and mules on the farms of America is because machines have not been versatile enough.

There can be no compromise between the machine and the horse. The machine that does not completely eliminate the necessity for horses on any given farm fails completely to meet the issue of cost.

The Ford Tractor with Ferguson System was designed to eliminate the horse. Not to *supplement*, but to *eliminate* him because he is a waster of land and time, the primary wealth of the farmer.

A generation of tractor experience has proved the horse cannot be eliminated simply by substituting a good machine in front of the same old implements.

So we threw overboard all the old ideas about pulling and controlling an implement in the soil, and developed a new principle of applying power. This principle is so efficient that this light tractor, using little fuel, will not only pull tools for which a heavy tractor has heretofore been necessary, but will go everywhere, and do everything the horse will do, and do it better.

The Ford Tractor with Ferguson System is not just another tractor. It is a new SYSTEM of land cultivation.

It is the beginning of a new era of low-cost production on the farm.

Ford Tractor FERGUSON SYSTEM

© Ford Motor Co.

The Ford Tractor with Ferguson System is sold nationally by the Ferguson-Sherman Manufacturing Corporation, Dearborn, Michigan, and distributed through dealers in every part of the country.

Only the Ferguson's versatility can surpass that of the horse. So now eliminate them because they waste land and time.

10 FARM JOURNAL and FARMER'S WIFE • MAY 1942 • FARM JOURNAL and FARMER'S WIFE 11

STEEL WHEELS

Ford Tractors with Ferguson System are now available on steel wheels. These Tractors have been designed to give outstanding performance and economical service on this new wheel equipment.

White line indicates extra size and weight of other equipment of equal working capacity.

100,000 tons of metal saved for armament *by this tractor!*

Since the Ford Tractor with Ferguson System was introduced, 2½ years ago, more than 100,000 units have been put to work on farms.

Each Ford Tractor, with an average complement of Ferguson Implements, represents a *clear saving of more than a ton of needed metal*, as compared with any other farm equipment of equal working capacity. 100,000 tons of steel saved for war production!

That is because the exclusive Ferguson System makes built-in extra weight unnecessary, either for traction or to get and hold soil penetration.

In the Ferguson System, tractor and implement are joined by linkage to operate as a single unit, under finger tip hydraulic control. It builds up traction—pulling power—just as needed, without extra built-in weight, and does it automatically.

No manual strength is needed, to raise and lower the implement, or control its depth in the soil, or to change from one implement to another. Saves time—saves operating cost.

The Ford Tractor with Ferguson System, with 2-plow capacity, is as easy to run as a modern automobile—can be efficiently and safely operated by women, elderly people, or young boys and girls. It gets farm work done on time—even when you are shorthanded—and keeps farm production costs down.

Any Ford Tractor dealer will gladly prove this for you, in a demonstration on your own farm, under *your* farming conditions.

© Ford Motor Co.

100,000 TONS IS STEEL ENOUGH TO BUILD

4 average battleships
or 48 average destroyers
or 5,500 medium tanks
or 23,000 scout cars
or 400,000 anti-aircraft guns with mounts
or more than 20,000,000 automatic rifles

Delivery subject to Government regulation of critical materials.

Ferguson-Sherman Mfg. Corp.,
Dearborn, Mich., Dept. 15

( ) Please send me more information
( ) Please arrange for a demonstration

Name________________

Address________________

...and this lighter weight saves *millions of gallons of fuel*

"We can flat-break eight acres, six inches deep, in ten hours on nine gallons of fuel." *J. V. and T. F. Holloway, Falcon, Alabama.*

J. W. Mack, Watannah, Indiana, plants fifteen acres of corn per day on six gallons of fuel—4/10 of a gallon per acre.

"Plows an acre an hour on a gallon"—with two 14" bottoms in heavy going—that's what thousands of owners report on the Ford Tractor with Ferguson System. For light draft work—cultivating, planting, mowing—the fuel consumption is *much* lower.

Here is the reason for this unusual economy. Because the Ford Tractor weighs less, it does not waste fuel on light draft work. Because of the Ferguson System of hydraulic control, lighter, weight-saving implements are thoroughly practical. These unit implements, with no friction or rolling resistance, use much less fuel. The Ford Tractor with Ferguson System saves metal, saves rubber, saves human energy, and saves fuel—millions of gallons when fuel is urgently needed for war.

L. H. Sparkman, Shamrock, Texas, writes, "I cultivate fifty acres of cotton on ten gallons of gas. My upkeep is nothing."

F. H. Cloud, Riverside, California, using tiller in orange grove covers ten-acre block two ways in eight to ten hours. One gallon an hour or less.

S. P. Munn, Bertram, Texas, reports 690 hours operation including plowing and discing, on 444 gallons of fuel. Total operating cost of six cents per hour.

•

Large agricultural college operated Ford Tractor 500 hours on general farm work, using 303 gallons of gas—6/10 of a gallon per hour.

•

Ohio farmer writes: "In plowing on our 500-acre farm this spring I used eleven gallons of gas in the day's work while my father used about 17 gallons of fuel oil in another make of tractor. I pulled a two-bottom 14" plow and he pulled a two-bottom 12" plow and I would gain three rounds a day on him in the Ford."

•

A. W. Nylund, Everett, Wash., says: "My farm is heavy, river bottom, clay loam soil. With my two-bottom 14" plow I average a gallon of gas an hour turning tough sod."

L. T. Cox, Alva, Oklahoma, writes, "We cut a crop of Austrian peas with Ford Tractor pulling combine in seven hours on five gallons of fuel."

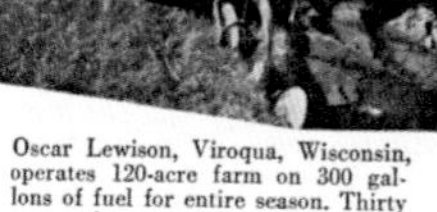

Oscar Lewison, Viroqua, Wisconsin, operates 120-acre farm on 300 gallons of fuel for entire season. Thirty acres of custom plowing included.

These are not "professional tests"—they are results secured by farmers under actual field conditions.
We are so confident of the economy and usefulness of this equipment—of its value to our country today—that we are willing to prove it on your farm against any competition.

FERGUSON-SHERMAN MFG. CORP.
DEARBORN, MICHIGAN

"I'm in the tractor seat eight or nine hours a day and use about five gallons of gas daily when cultivating nursery shrubbery." *Richard Sliger, Painesville, Ohio.*

Ferguson saves steel for the war time effort. Yet steel wheels are offered because there was a wartime shortage of rubber! May 1942.

A tractor for four seasons, which can be operated by boys, or for that matter, girls!

The tractor on this farm spends the nights in the barn. But not the days, either winter or summer. For it is a working tractor—working four full seasons a year—wherever there is work to be done. It's a Ford Tractor with Ferguson System. It doesn't demand a winter vacation.

On your farm, winter may not bring cold and snow. But still the farm year is a twelve-month year, especially if you're going to handle those "when-I-get-time" jobs. You can make the season between crops mighty profitable, if you have the right equipment.

There are lots of good tractors made nowadays. Many of them are three-wheel jobs—cultivating machines. They do good work, but they're specialists. Big trouble is, they are not easily suited to do *all* the work on your farm.

Compare one of those machines with the Ford Tractor with Ferguson System.

It won't take you long to see the advantage of four-wheel stability, of automobile type easy steering, of quick adaptability to one job after another, with implements changed in a minute or less.

Then, there is the Ferguson System—a new method of attaching implements and controlling them in the ground—making implement and tractor a complete, easy-to-operate unit.

That makes this lightweight, economical tractor a tough competitor in the middleweight class, without asking you to buy extra weight, or drag it around the farm.

Pull two 14-inch plows, or a big two-gang disc, cultivate a section or a kitchen garden, mow heavy crops, break new ground—it's all the same to the Ford Tractor with Ferguson System. The System does it—makes this tractor different from all others—gets more work out of a gallon of gas than you ever thought possible.

It will help you get all of your regular farm planting and cultivating and harvesting done *on time*. That's important. And then it's ready to do the off-season jobs—clearing, cutting wood, filling silos, grinding feed, pumping water, farm hauling—whatever you want done.

This four-wheel, four-season equipment is ready to go at the touch of the starter, lets you off easy on gas and oil, and asks no questions about the kind of job you want it to do next. The boys can run it, too, or for that matter, the girls.

Sounds like a big contract, but you'll believe it when you see a demonstration, right on your own farm. The nearest Ford Tractor Dealer will gladly take care of that for you.

© Ford Motor Co.

The Ford Tractor with Ferguson System is sold nationally by the Ferguson-Sherman Manufacturing Corporation, Dearborn, Michigan, and distributed through dealers in every part of the country.

Wood Bros. were approved Ferguson equipment suppliers for use with Ferguson tractors. August 1944.

THE
FERGUSON SYSTEM
..as advanced as jet propulsion
The Ferguson System turned the tractor into a farming machine
One of MANY advantages
WITH THE FERGUSON SYSTEM
EACH ATTACHED IMPLEMENT BECOMES SELF-PROPELLED AND AUTOMATICALLY-CONTROLLED
50
SUCCESSFUL FARMING, OCTOBER, 1945

1 SAVE VITAL CROP-MAKING TIME. You can change from one crop-making or crop-saving job to another, in a minute or less. With the Ferguson System you just insert three pins to attach an implement. It's as simple and quick as stirring a cup of coffee.

2 DO BACKBREAKING JOBS THE EASY WAY. With the Ferguson System you can put your plow into a fence corner with only a touch on the finger tip control lever. Use all your land and save your back, too.

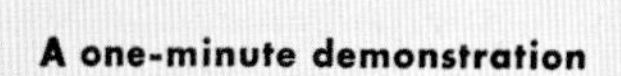

of BETTER LIVING

BY FERGUSON SYSTEM

This one-minute demonstration shows how the Ferguson System can bring Better Living to your farm. It shows how tasks are made simpler—that saves your time. It shows how jobs are made easier—that saves your strength. By using your management ability, you can turn both into Better Living.

3 YOUNG BOYS DO TOUGHEST WORK SAFELY. Two big advantages of the Ferguson System make this possible—one, the front end is held down in toughest going; two, the hydraulic lift raises and lowers implements—does away with long, knuckle-skinning levers.

4 MAKE MORE MONEY—HAVE TIME TO ENJOY IT. When you do your regular farm work faster and easier, you have more time and energy for more profit-making opportunities—and time to do things with your family.

Ask for a demonstration on your own farm!

There's a new farming experience awaiting you and your family the day you try the Ferguson System on your place. Just the feel of how easily and simply it operates will tell you volumes about how much more work you will be able to do. You will see how it takes the drudgery out of farm work and how self-propelled and automatically-controlled implements do more in a day than you ever thought possible.

For your own satisfaction, get the feel of the Ferguson System firsthand—ask your friendly Ferguson Dealer for a demonstration.

HARRY FERGUSON, INC. • DETROIT, MICH.

SUCCESSFUL FARMING, OCTOBER, 1945 51

A new post-war twist to an old message – "The Ferguson System turned the Tractor into a Farming Machine". We now see Harry Ferguson, Inc. emerging as the Ferguson representation. This is an early example of the use of blue colour in Ferguson USA advertising. Be as modern as a jet plane. October 1945.

COUNTRY GENTLEMAN
December, 1945
AS ADVANCED AS RADAR . . .
BETTER LIVING
A one-minute demonstration of the
FERGUSON SYSTEM
1
QUICK, EASY WHEEL SPACING. Makes possible a row crop tractor and a four-wheel tractor in one. Spacings made without special equipment, and without altering steering—an exclusive Ferguson System advantage.
One of MANY advantages . . THE FERGUSON SYSTEM automatically protects implements when striking an obstacle

More comparisons with advanced technology – this time radar. The essence of which can be demonstrated in a minute. December 1945.

Your Post-War Tractor should . .
AUTOMATICALLY CHANGE its "WEIGHT" to SUIT the JOB
Only the FORD TRACTOR with FERGUSON SYSTEM gives you this ECONOMY – today
WHEN YOU THINK about your tractor of the future, you have a right to expect that it will give you freedom from guessing how much weight to pile on to pull through those tough spots on your farm. And freedom from straining your back, skinning your knuckles and barking your shins putting cumbersome steel or concrete wheel weights into place.
It also should be unnecessary for you to pay for the hundreds of pounds of dead metal built into ordinary tractors of today, plus the fuel to carry it around, merely to provide enough traction to do an ordinary day's work.
These things you have a right to expect — why? Because the Ford Tractor with Ferguson System gives them to you today. It is the only tractor which automatically changes its "weight" to suit whatever job you are doing.
Downward pressure exerted on the tractor through the patented and exclusive Ferguson System replaces the excess built-in weight and added wheel weights necessary to other tractors.
As the two bottom links pull, the upper link is pushed forward and downward by the effort of the implement to tip in the direction of pull.
When the draft gets heavier, so does the downward pressure through the top link, and the tractor automatically provides itself with extra traction to do the job.
Thus, the Ford Tractor with Ferguson System gives you an amazing degree of economy no other tractor can offer — a light tractor for light work and a "heavier" tractor when more traction is needed.
This is only one of the many patented and exclusive features of this revolutionary tractor. The next time you drive by your Ford-Ferguson Dealer, stop in. He will show you all the reasons why this advanced farming machine drives crop production costs to an all-time new low level.
HARRY FERGUSON, INC.
Dearborn, Michigan
Ford Tractor
FERGUSON SYSTEM

The weight transfer message subtly changed to changing the weight of the tractor.

Your Post-War Tractor should . .
have a Mechanical Brain and Automatic Muscles of Steel to
SAVE YOUR BACK and ARMS
Only the FORD TRACTOR with FERGUSON SYSTEM gives you this greater OPERATING COMFORT today
AS YOU LOOK to the future, you have good reason to expect that your new tractor will take over an even greater share of the hard back-and-arm labor of farming.
The tractor of the future should automatically control the depth at which implements operate. Bruised knuckles, pinched fingers, backache and tired arms from frequent adjustments of plow or cultivator with heavy levers ought to be only unpleasant memories.
You have good reason to expect this greater operating comfort, because the Ford Tractor with Ferguson System gives it to you today. It is the only tractor with a mechanical brain and automatic muscles of steel to save your back and arms.
The touch of your finger tips on a five-inch control lever directs a mechanical brain which operates automatic muscles of steel to set the implement at the depth you desire.
The tractor then actually "thinks" and acts for itself, as the patented and exclusive Ferguson System automatically maintains constant control over the depth of operation—hard manual labor of reaching and pulling that formerly took its daily toll of your back and arms.
Around the barnyard, too, these automatic muscles of steel come in mighty handy. All you use is the strength in your finger tips to set them in motion, and they will do such back-breaking jobs as dipping a hog or raising the corner of a sagging hen house.
Saving your strength to enjoy the profits from your farm and life with your family is only one of the many advantages of this amazing tractor.
See your Ferguson Dealer at your earliest opportunity. He will be glad to show you all of the ways this advanced tractor is revolutionizing life on the farm.
Ford Tractor
FERGUSON SYSTEM
The only Tractor that Automatically Changes its 'Weight' to Suit the Job.
HARRY FERGUSON INC., Dearborn, Michigan

The post-war tractor must have a brain. A real challenge to other manufacturers.

COUNTRY GENTLEMAN October, 1946

* THE FERGUSON SYSTEM

AS FAR AHEAD AS THE HELICOPTER

All farm machinery saves time, but some saves much more time than others.

For example, one thing that makes the Ferguson System such an amazing time saver is that it enables you to attach implements in only one minute. Even a boy can attach a Ferguson Implement without heavy straining or lifting. During busy seasons he can change implements—shift from one job to another—without delay, to get the work done at the best time, for best results. And how he will love it! What better way for a young person to become interested in farming than to learn firsthand that farm life can be modern, enjoyable and profitable?

Not just one Ferguson Implement has this quick-change advantage. You get it right through all of the Ferguson self-propelled and automatically-controlled implements. Let us demonstrate this on your farm, with the other Ferguson System advantages pictured on the opposite page, and many more.

Ten minutes behind the wheel will convince you that the way to better farming and better living is through the Ferguson System. It uses hydraulic power instead of muscle power. It sets the farmer free.

Ask your FERGUSON DEALER for a demonstration on your farm

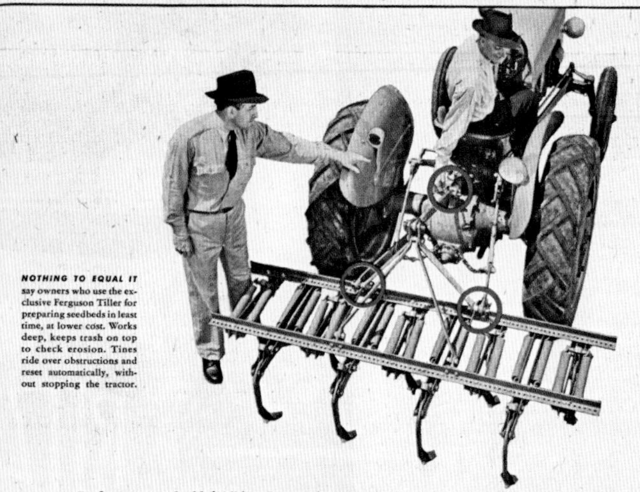

NOTHING TO EQUAL IT say owners who use the exclusive Ferguson Tiller for preparing seedbeds in least time, at lower cost. Works deep, keeps trash on top to check erosion. Tines ride over obstructions and reset automatically, without stopping the tractor.

Only one of *MANY* advantages . . . THE FERGUSON SYSTEM provides *ONE MINUTE - 3 POINT* attaching and detaching of implements

COUNTRY GENTLEMAN

1 FINGER TIP CONTROL LIFTS TRACTOR OFF ALL FOUR WHEELS—With the Ferguson System this light, handy jack is slipped under the tractor and quickly attached to the Ferguson linkage. Then a touch of the finger tip control lifts the tractor off all four wheels at once, for easy changing of tires or tread widths.

2 ONE IMPLEMENT, MANY USES—The Ferguson Spring Tine Cultivator is easily and quickly converted from summer row-crop work to general field cultivation during the fall and spring. Like other Ferguson Implements, it is of special alloy steel construction that brings you a new freedom from repair costs.

3 FERGUSON TERRACER AND UTILITY BLADE—Builds terraces, digs or cleans ditches, grades roads, cleans feedlots and feeding floors, clears snow; literally a dozen uses on every farm for your own work and profitable custom jobs. Attached or detached in a minute or less. Blade tilts 45 degrees up or down, throws to right or left.

4 EXPAND FARMING OPERATIONS WITH NO MORE WORK—The time and strength saved through each day by using the Ferguson System adds up. This time can be used for additional, money-making farm activities such as keeping more cows, as well as for those hours of recreation and rest that mean so much to the modern farmer.

*** THE FERGUSON SYSTEM turned the tractor into a farming machine**

- Uses natural laws instead of heavy inbuilt weight to gain penetration and automatically adjust traction to suit the job.
- Enables you to lift, lower, set and control implements by hydraulic power instead of muscle power.
- Aids profitable, flexible farm operation by one-minute implement attaching, one-wrench adjustments and easy adaptability to the widest range of farm work.
- Provides automatic protection against hidden obstacles without "losing" the implement.

HARRY FERGUSON, INC., DETROIT, MICHIGAN

This time the Ferguson is ranked with helicopter technology. One minute to hitch tractor to implement, ten minutes to convince yourself that you need the Ferguson System. October 1946.

AS MODERN AS TELEVISION

Yet another comparison with other modern technology of the day – the television. Repeated emphasis on weight transfer from implement to tractor. March 1946.

Only five advertisements were found for the Ford Ferguson tractor in 1947 as it was their last year of production. Harry Ferguson was already planning production of his own tractor at Detroit in a new factory following the break up of relationships with Ford. The tiller was one of the original types of implements sold with the Ferguson Brown and is seen here surviving to the end of the second generation of Ferguson tractors.

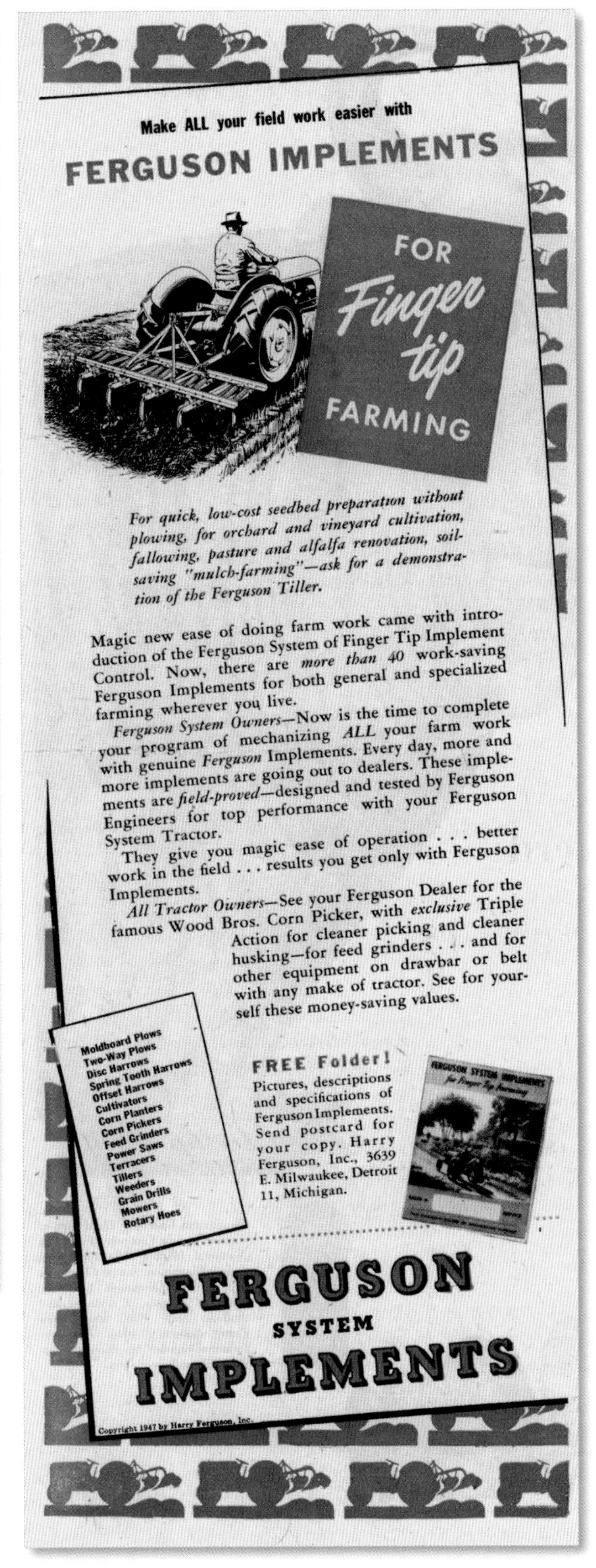

Make ALL your field work easier with
FERGUSON IMPLEMENTS
FOR Finger tip FARMING
For quick, low-cost seedbed preparation without plowing, for orchard and vineyard cultivation, soil-fallowing, pasture and alfalfa renovation, soil-saving "mulch-farming"—ask for a demonstration of the Ferguson Tiller.
Magic new ease of doing farm work came with introduction of the Ferguson System of Finger Tip Implement Control. Now, there are more than 40 work-saving Ferguson Implements for both general and specialized farming wherever you live.
Ferguson System Owners—Now is the time to complete your program of mechanizing ALL your farm work with genuine Ferguson Implements. Every day, more and more implements are going out to dealers. These implements are field-proved—designed and tested by Ferguson Engineers for top performance with your Ferguson System Tractor.
They give you magic ease of operation . . . better work in the field . . . results you get only with Ferguson Implements.
All Tractor Owners—See your Ferguson Dealer for the famous Wood Bros. Corn Picker, with exclusive Triple Action for cleaner picking and cleaner husking—for feed grinders . . . and for other equipment on drawbar or belt with any make of tractor. See for yourself these money-saving values.
Moldboard Plows
Two-Way Plows
Disc Harrows
Spring Tooth Harrows
Offset Harrows
Cultivators
Corn Planters
Corn Pickers
Feed Grinders
Power Saws
Terracers
Tillers
Weeders
Grain Drills
Mowers
Rotary Hoes
FREE Folder!
Pictures, descriptions and specifications of Ferguson Implements. Send postcard for your copy. Harry Ferguson, Inc., 3639 E. Milwaukee, Detroit 11, Michigan.
FERGUSON SYSTEM IMPLEMENTS
Copyright 1947 by Harry Ferguson, Inc.

RAISES...LOWERS
THROWS IN OR OUT OF GEAR

You have complete working control of this new 13 x 7" Dempster Single Disc Grain Drill for the Ferguson System—right from the tractor seat. With Finger Tip Control, you lower or raise the disc openers for turning or transport. Drill is automatically thrown out of gear when raised—goes in gear when lowered.
You adjust seeding depth in an instant—½ to 4 inches deep—or you can do surface seeding just as simply.
Exclusive New Feed
Will Meter Seed Accurately
At last, a new basic improvement in grain drill design! It's the adjustable internal run feed of this new grain drill, available through Dealers selling Ferguson Implements.
As illustrated, width of feed rings may be adjusted instantly for accurate drilling, both of size and quantity, of any seed from alfalfa to soybeans. You change no sprockets or gear ratios—merely shift two levers on plainly marked seed index plates...See your Ferguson Implement Dealer for a demonstration.

Grader & Terracer

THE MEIKLEJOHN CO.
198 North Main Street
Fond du Lac, Wisconsin

No need for a man on the drill with the new Ferguson seed drill which went on to be a very successful implement in the next generation of Ferguson tractors. 1947.

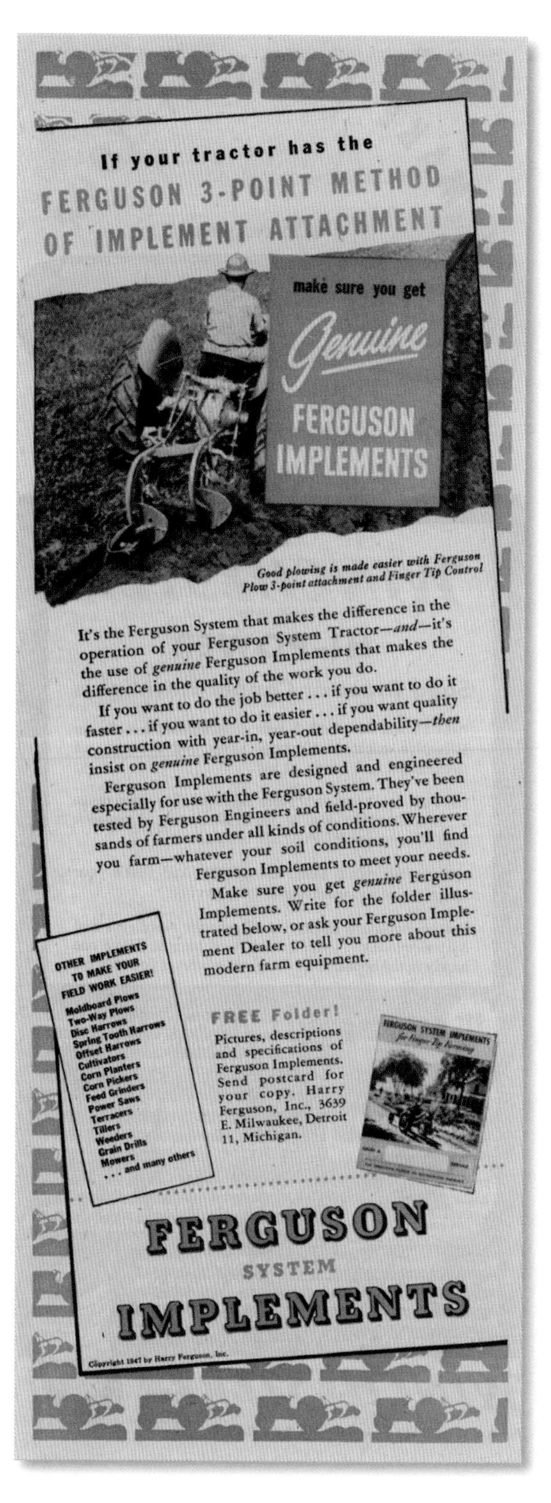

If your tractor has the
FERGUSON 3-POINT METHOD OF IMPLEMENT ATTACHMENT
make sure you get
Genuine
FERGUSON IMPLEMENTS
Good plowing is made easier with Ferguson Plow 3-point attachment and Finger Tip Control
It's the Ferguson System that makes the difference in the operation of your Ferguson System Tractor—and—it's the use of genuine Ferguson Implements that makes the difference in the quality of the work you do.
If you want to do the job better...if you want to do it faster...if you want to do it easier...if you want quality construction with year-in, year-out dependability—then insist on genuine Ferguson Implements.
Ferguson Implements are designed and engineered especially for use with the Ferguson System. They've been tested by Ferguson Engineers and field-proved by thousands of farmers under all kinds of conditions. Wherever you farm—whatever your soil conditions, you'll find Ferguson Implements to meet your needs.
Make sure you get genuine Ferguson Implements. Write for the folder illustrated below, or ask your Ferguson Implement Dealer to tell you more about this modern farm equipment.
OTHER IMPLEMENTS TO MAKE YOUR FIELD WORK EASIER!
Moldboard Plows
Two-Way Plows
Disc Harrows
Spring Tooth Harrows
Offset Harrows
Cultivators
Corn Planters
Corn Pickers
Feed Grinders
Power Saws
Terracers
Tillers
Weeders
Grain Drills
Mowers
... and many others
FREE Folder!
Pictures, descriptions and specifications of Ferguson Implements. Send postcard for your copy. Harry Ferguson, Inc., 3639 E. Milwaukee, Detroit 11, Michigan.
FERGUSON
SYSTEM
IMPLEMENTS
Copyright 1947 by Harry Ferguson, Inc.

The basic concept of the Ferguson two furrow plough has now survived some 30 years and is ready to go forward to the TE and TO generations of tractors – and even into the later Massey Ferguson era. 1947.

MUSCLE-SAVING...TIME-SAVING
Finger tip Farming
for ALL your work
WITH
Genuine
FERGUSON IMPLEMENTS
For better, easier discing, there's the Ferguson Tandem Disc Harrow. Gangs are angled or straightened by Finger Tip Control without stopping the tractor.
There's a world of difference in implements . . . a big difference in the quality of work they do.
It's only natural that owners of Ferguson System Tractors should expect—and get—best performance with genuine Ferguson Implements. These implements have been engineered and field-proved to help you get the most out of the Ferguson System. They help do all your work easier, at rock-bottom cost.
Now, there are more than 40 work-saving, time-saving Ferguson Implements for both general and specialized farming. More and more of them are being shipped to dealers every day.
Owners of other tractors—See the famous Wood Bros. Corn Picker, Feed Grinders, Power Saws and other implements that work with any tractor.
Write for the FREE FOLDER illustrated below, or ask your Ferguson Implement Dealer for a demonstration of the implements you need.
FERGUSON IMPLEMENTS TO MAKE YOUR FIELD WORK EASIER!
Moldboard Plows
Two-Way Plows
Disc Harrows
Spring Tooth Harrows
Offset Harrows
Cultivators
Corn Planters
Corn Pickers
Feed Grinders
Power Saws
Terracers
Tillers
Weeders
Grain Drills
Mowers
Rotary Hoes
FREE Folder!
Pictures, descriptions and specifications of Ferguson Implements. Send postcard for your copy. Harry Ferguson, Inc., 3639 E. Milwaukee, Detroit 11, Michigan.
FERGUSON
SYSTEM
IMPLEMENTS

The trailed disc harrow served the Ford Ferguson era well, and carried over for a while into the TE 20 and TO 20 generations of tractors until the mounted disc harrow was developed.

Possibly one of the last Ford Ferguson advertisements, and in a style much adopted by the TO tractor era in the following year. March 1947.

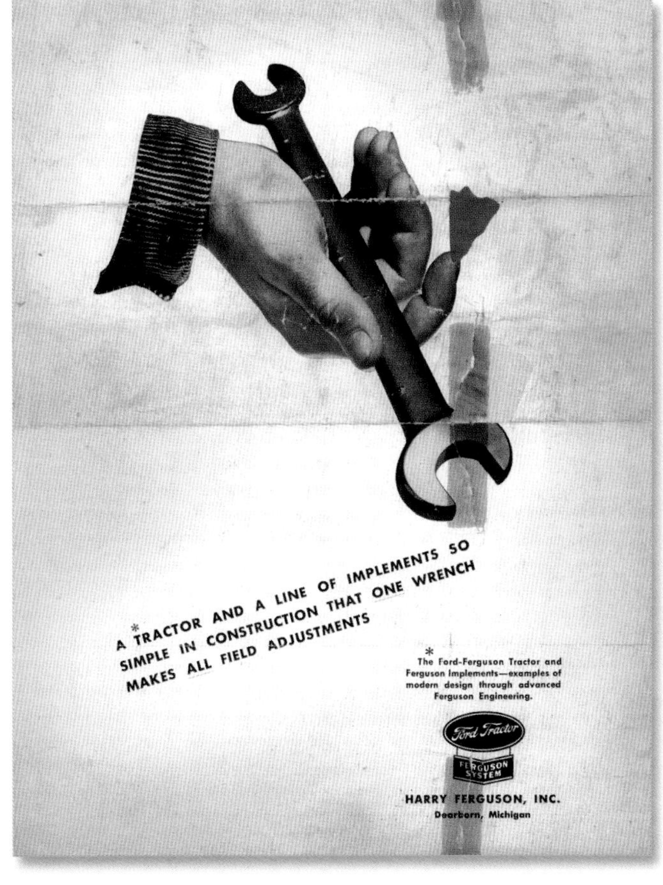

Harry Ferguson's famous spanner. This spanner was issued with all tractors from the Ferguson Brown era to those of the MF 35 and 65 era. They are now keenly sought after collectors' items with several versions to be found.

THE FERGUSON TE20

Following the break with Ford in 1947 and consequent cessation of Ford Ferguson tractor production, it was lucky that Harry Ferguson had been looking at how to manufacture Ferguson tractors in the UK as far back as 1943. Ford had determined not to make a Ferguson System tractor in the UK. Sir John Black, owner of the Standard Motor company had an empty factory at Banner Lane in Coventry. They signed an agreement for him to manufacture Ferguson TE tractors for Harry Ferguson in late 1946. Ferguson had also had his design team working on this new tractor and a comprehensive implement range since 1945 and Harry Ferguson progressively moved his design team to another of Black's unused facilities on Fletchampstead Highway not far from Banner Lane. The TE Ferguson tractor was very much a revised Ford Ferguson and notably had, eventually, a Standard overhead-valve engine to replace the Ford side-valve unit. It was more powerful and a little heavier. The tractor was also given four forward gears in contrast to the Ford Ferguson's three. Of all the Ferguson tractors made the TE20s had the greatest range of models – about 25 – which included various chassis and engine combinations. The various chassis facilitated narrow, orchard and vineyard tractors but by far the greater number of tractors produced were standard agriculturals.

So again Harry Ferguson had someone else manufacturing his tractors and although he designed his implements these too were made by outside manufacturers. The range of tractors also included industrials for which there were a range of Ferguson "Approved" implements.

The TE20 tractor started life with a USA built Continental engine due to an initial shortage of UK engines but this was soon replaced by a Standard 80mm engine. This was subsequently replaced by a slightly more powerful 85 mm engine at serial number 172501, and a Standard diesel engine (12V) option was eventually introduced in March 1951 at serial number 200,001. 12V electrics then came in at serial number 250001 for all petrol and TVO tractors.

NEW HOPE FOR A HUNGRY BRITAIN
To-day, British farmers need—and deserve—nothing less than what has been called "the most efficient farming machine the world has ever seen"—the Ferguson Tractor with Ferguson Implements.
Equipped with this major revolution in agricultural engineering, they will get more work done, better, faster; with greater efficiency and less effort than has ever before been possible. That is why, in the name "Ferguson" British farmers see new hope for a hungry land.
The Banner Lane factory of The Standard Motor Co. Ltd., Coventry is exclusively devoted to the manufacture of Ferguson Tractors for
Harry Ferguson, Ltd.
COVENTRY
ASK YOUR DEALER FOR A DEMONSTRATION ON YOUR FARM
FERGUSON SYSTEM

Despite starting production of tractors in the UK in 1946, it seems that serious advertising did not start until 1948, and even then initially some were only half page advertisements. This first one in January plays on the post-war hunger of UK when food supplies were still very limited. January 1948.

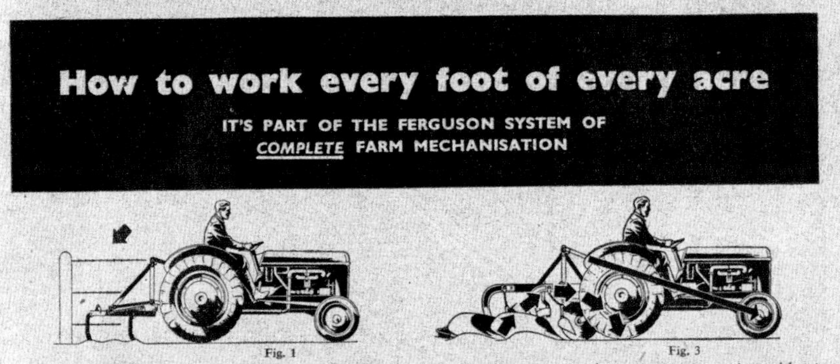

Cultivate every foot of every acre by working where other implements won't. Work within inches of obstructions, walls, hedgerows, and produce more food at less cost. You *can* with the Ferguson System of *complete* farm mechanisation. 3-point linkage makes this possible by coordinating action of tractor and implement, making them work as a single unit. Tractor carries implement quickly into position. (*Fig. 1*) Finger-tip hydraulic control from the driver's seat (*Fig. 2*) lowers implement exactly where you want to start work, automatically maintains working depth of implement. You can work close to fences and in confined areas. You can farm from the driver's seat.

Fig. 2

The Ferguson System gives you more traction without built-in weight that packs the soil, impedes drainage. 3-point linkage and hydraulic depth control automatically adjust weight on tractor's rear wheels (*Fig. 3*) to provide necessary traction and keep front end down.

ASK YOUR FERGUSON DEALER FOR A DEMONSTRATION ON YOUR FARM

**Every week more than 1,000 Ferguson Tractors are produced in the Banner Lane factory of the Standard Motor Company Ltd., and more than 5,000 Implements by the foremost British manufacturers, for Harry Ferguson Ltd., Coventry.*

FARM BETTER FARM FASTER WITH **Ferguson** FERGUSON SYSTEM

S2A

Get a demonstration! Production now up to 1000 tractors per week and 5000 implements. May 1948.

Ferguson Hydraulic Jack Makes Wheel Spacing Easy

TRACTOR LIFTS ITSELF! With the aid of a "Ferguson" hydraulic jack, one man can easily change, replace or offset "Ferguson" tractor wheels as desired. Simply place jack in position under front and rear axle of tractor. A touch of the hydraulic control lever lifts all four wheels of tractor off the ground. No lifting, placing or levering difficulties. It's part of the "Ferguson System" of *complete* farm mechanisation. Ask your "Ferguson" Dealer for a demonstration on your farm.

FERGUSON SYSTEM **FARM BETTER, FARM FASTER WITH FERGUSON**

"Ferguson" Tractors are manufactured by The Standard Motor Company Ltd. for Harry Ferguson Ltd., Coventry S15A

The Ferguson jack for all four wheels at the same time completed the Ferguson farm mechanisation system. September 1948.

Swedish Test Fails to Break Hydraulic Lift

49,000 TIMES! Swedish agricultural engineers recently subjected the hydraulic lifting mechanism of the "Ferguson" tractor to the toughest test on record. A "Ferguson" transport box with a 5 cwt. load was continuously lifted and lowered 49,000 times without impairing the pump's efficiency. Finger-tip hydraulic *and* automatic hydraulic control of implements carried by the tractor are both part of the "Ferguson System" of *complete* farm mechanisation. Ask your "Ferguson" Dealer for a demonstration on your farm.

FERGUSON SYSTEM

Farm Better, Farm Faster with FERGUSON

"Ferguson" Tractors are manufactured by The Standard Motor Company Ltd., for Harry Ferguson Ltd., Coventry S.16A

The Swedes traditionally produced some of the world's finest steel, but the Ferguson hydraulic lift withstood even their engineers' tests. October 1948.

QUICK CONVERSION! The simple attachment of a third bottom to your existing "Ferguson" 10" 2-furrow plough increases ploughing capacity by over 50% and gives more traction in slippery conditions. 3-point linkage between "Ferguson" tractor and plough, and hydraulic control of the plough, eliminate the need of wheels, axles, levers, springs or rope-operated clutches. It's part of the "Ferguson System" of *complete* farm mechanisation. Ask your "Ferguson" Dealer for a demonstration on your farm.

FERGUSON SYSTEM

Farm Better, Farm Faster with FERGUSON

"Ferguson" Tractors are manufactured by The Standard Motor Co. Ltd. for Harry Ferguson Ltd., Coventry 17A

The plough could be increased in size if needed. 1948.

How to work every foot of every acre

IT'S PART OF THE FERGUSON SYSTEM OF COMPLETE FARM MECHANISATION

Work within inches of obstructions, walls, hedgerows. Produce more food at less cost. You *can* with the Ferguson System of *complete* farm mechanisation. 3-point linkage makes tractor and implement work as a single unit. Tractor carries implement quickly into position.

Finger-tip hydraulic control from the driver's seat lowers implement exactly where you want to start work, automatically maintains working depth of implement. You can work close to fences and in confined areas. You can farm from the driver's seat.

Ask your Ferguson Dealer for a demonstration on YOUR farm.

**Every week more than 1,000 Ferguson Tractors are produced in the Banner Lane factory of the Standard Motor Co. Ltd., and more than 5,000 Implements by the foremost British manufacturers, for Harry Ferguson Ltd., Coventry.*

FERGUSON SYSTEM

FARM BETTER, FARM FASTER WITH

Ferguson

82B

Still the same message from Ferguson Brown days – every inch of land is cultivable with a Ferguson. April 1948.

APRIL 1, 1948 FARM IMPLEMENT AND MACHINERY REVIEW 1127

This is the famous Ferguson Tractor

The Ferguson tractor is one of Britain's leading exports. A $20,000,000 American order, a £2,000,000 French order and other orders from Europe and the Commonwealth are already bringing £7,000,000 to Britain.

The Ferguson tractor, as part of the Ferguson System of *complete* farm mechanisation, is in world-wide demand by farmers because it enables them to produce *more* food at *less* cost from every available acre.

**Every week more than 1,000 Ferguson tractors are produced in the Banner Lane factory of Standard Motor Co. Ltd., and more than 5,000 Ferguson implements by foremost British manufacturers, for Harry Ferguson Ltd., Coventry.*

FARM BETTER
FARM FASTER WITH **Ferguson**

Just look at those export figures – and less than two years after production started at Coventry, England. April 1948.

Possibly one of the first uses of colour advertising by Ferguson in the UK. The Ferguson's capacity for haulage is the new theme – note the trailer has the early hitch design which is now very rare. September 1948.

THE FARMERS WEEKLY, SEPTEMBER 24, 1948

WAGES: WORKER'S CASE FOR REVISION

THE FARMERS WEEKLY

VOL. XXIX. No. 13 (REGISTERED AT THE G.P.O. AS A NEWSPAPER) SIXPENCE

Why Farmers Prefer New Ferguson Trailer

SAVE HAULING TIME! You can with the new "Ferguson" 3-ton hydraulic tipping trailer, now ready for delivery. Tractor and trailer are easily connected by "Ferguson" patent attachment. 1/3 of the trailer load's weight is automatically transferred forward of the tractor's rear axle to give more traction, more front-wheel stability.

UP IT GOES! Only with the "Ferguson" tractor-trailer unit can you elevate trailer load by finger-tip hydraulic lift control from the driver's seat. Maximum tipping position in less than 2 minutes. Intermediate tipping positions also controlled from driver's seat. Ask your "Ferguson" Dealer for a demonstration on your farm.

OFF TO MARKET! Like every other farmer, a very considerable part of your time is probably spent hauling. With the new "Ferguson" trailer, as part of the "Ferguson System" of *complete* farm mechanisation, you can save valuable hauling time by working faster and more efficiently. You can work where other trailers won't budge.

FERGUSON SYSTEM

Farm Better, Farm Faster with FERGUSON

"FERGUSON" TRACTORS ARE MANUFACTURED BY THE STANDARD MOTOR COMPANY LTD., FOR HARRY FERGUSON LTD., COVENTRY

44 *The Farmers Weekly, November 12, 1948*

Yorkshire Farmer Ploughs Motorcycle Test Hill

GRADIENT OF 1 IN 2½! This hill was ploughed for the first time by a "Ferguson" tractor with a 10" 2-furrow plough working to a depth of 8". The gradient is 1 in 2½! Working on gradients or hillsides is easy and safe for the "Ferguson" because it automatically adjusts its weight to the job. 3-point linkage and hydraulic control co-ordinate action of tractor and implement. You get more traction, more front wheel stability, without built-in weight. It's part of the "Ferguson System" of *complete* farm mechanisation. Ask your "Ferguson" Dealer for a demonstration on your farm.

Farm Better, Farm Faster with FERGUSON

Another first for Ferguson. Ploughing up a hitherto unploughed motorcycle test hill. November 1948.

38 *The Farmers Weekly, December 10, 1946*

Ferguson Does These Essential Winter Jobs Better

WINTER PLOUGHING is easier with the "Ferguson System," even when the ground is hard. You get more traction without built-in weight because the "Ferguson" tractor automatically adjusts its weight to suit the job. Ploughing depth is maintained by automatic hydraulic control. You get better ploughing with lower costs and less fatigue.

FERGUSON SYSTEM

WHEN THE WEATHER makes working in the field impracticable, attach a "Ferguson" wood saw to the "Ferguson" tractor by simple 3-point linkage. Drive to the woodpile instead of hauling cumbersome uncut wood to the saw. You finish another essential farm job faster. It's part of the "Ferguson System" of *complete* farm mechanisation.

MUCK SPREADING, wood hauling, ballast carting take these jobs in stride with a "Ferguson" 3-ton hydraulic tipping trailer. Get more traction, work on slippery ground and where other trailers bog down. Tip the trailer in less than a minute by finger-tip hydraulic control. Ask your "Ferguson" Dealer for a demonstration on your farm.

Farm Better, Farm Faster with FERGUSON

"Ferguson" Tractors are manufactured by The Standard Motor Co. Ltd. for Harry Ferguson Ltd., Coventry

S 20

Jobs all winter for the Ferguson. But we need that muck spreader and muck loader as soon as possible! December 1948.

Ferguson Dealer Service Streamlined for Prompt Action

MOBILE SERVICE! Every "Ferguson" Dealer makes routine service calls at regular intervals on all owners of "Ferguson" tractors and implements in his territory, as well as emergency calls if required. Because of this exclusive on-your-farm service, all unnecessary delays are avoided. Part of your "Ferguson" Dealer's job is to make sure that you get maximum results from the "Ferguson System" of *complete* farm mechanisation. Ask him for a demonstration on your farm.

Farm Better, Farm Faster with FERGUSON

"Ferguson" Tractors are manufactured by The Standard Motor Co. Ltd. for Harry Ferguson Ltd., Coventry S19A

Harry Ferguson placed great emphasis on providing a high quality service to customers. This was essential as many farmers were buying and operating a tractor for the first time in this period of rapid changeover from horse to tractor on smaller farms. December 1948.

See the FERGUSON System of Farming at the ROYAL EASTER SHOW

BRITISH FARM EQUIPMENT COMPANY
(Division of Standard Cars, Ltd.)
STAND No. 109
Clydesdale Street

Ferguson Tractors and Full Range of Implements on Display.

Every Farmer should see this Ferguson System Exhibit

14 *March, 1948—POWER FARMING IN AUSTRALIA AND NEW ZEALAND*

The essence of the Ferguson tractor was the Ferguson System which brought together the integrated nature of tractor and implement design to achieve lightweight implements and tractors. This tractor is fitted with an American built Continental engine which was used to start production of the tractors in Coventry. March 1948.

Investigate the Ferguson System . . .

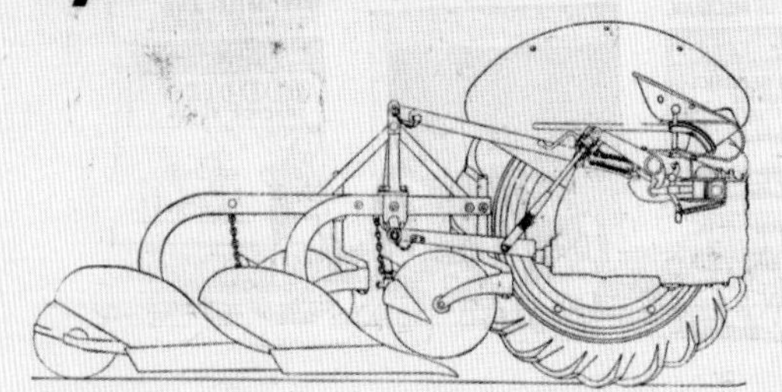

Sketch showing the Ferguson control gear and implement linkage. Note the convenient hydraulic control lever near driver's seat.

Implement Control . . . .

Each implement has one major control—the lift of the plough, the angle of the discs, the lift of the mower, etc.; and this control is so arranged that it can be operated with little effort on the part of the driver by means of a small lever working in conjunction with a hydraulic cylinder and pump attached to tractor. In the case of ploughs, cultivators, etc., the hydraulic control holds the implement at a pre-determined depth.

Three-Point Hitch . . . .

The most outstanding feature of the FERGUSON SYSTEM is the three-point hitch. As will be seen in the illustrations, the two lower hitch points are in the normal draw-bar position. The third point is above them. Thus, the pull of the implement exerts a **pushing** force above the point of draft. This spreads the load over all four wheels and keeps the front of the tractor down.

The dotted arrow shows how the plough's weight, plus the weight of the soil in the plough add weight to the tractor's rear wheels. Through the FERGUSON SYSTEM of 3-point linkage, the natural tendency of the implement to revolve as it is pulled into the ground is converted into a strong, forward-slant-thrust which holds the front wheels down also.

BIG SHIPMENTS CONSTANTLY ARRIVING — EARLY DELIVERY ASSURED

Ask your local Dealer — or write direct to

BRITISH FARM EQUIPMENT COMPANY

DIVISION OF STANDARD CARS LTD.

83-89 Flinders Street, Sydney 'Phone: FA 4183-4-5-6.

62 *April, 1948—POWER FARMING IN AUSTRALIA AND NEW ZEALAND*

Farmers had never previously experienced anything like the Ferguson "System" so it and the advantages had to be explained to them. April 1948.

far MORE than a tractor

..a completely NEW farming system that links Tractor and Implement as one unit.

The Ferguson System is one that employs natural laws instead of using heavy in-built weight to give traction and penetration. This ensures that the tractor automatically adjusts its weight to the job — that the implement maintains an even depth by automatic hydraulic control on all types of ground. With the Ferguson System only one model of tractor is required — and a light one at that — to suit every type of farm and kind of work.

PENETRATION WITHOUT WEIGHT

The weight of the implement, plus the weight of the soil on it, plus the drag of the implement automatically provide the traction required for the job. The Ferguson System makes it possible, for the first time in farming history, for a light-weight tractor to secure penetration without in-built dead weight.

THE FERGUSON SYSTEM

Ask your local dealer, or write to

BRITISH FARM EQUIPMENT CO.

(Division Standard Cars Ltd.)

83-97 FLINDERS ST., SYDNEY. 'PHONE: FA 4183

BIG SHIPMENTS CONSTANTLY ARRIVING EARLY DELIVERY ASSURED

N.S.W. DISTRIBUTORS:

BRITISH FARM EQUIPMENT CO.

83-97 FLINDERS STREET, SYDNEY

Address in Victoria: 568 Elizabeth Street, Melbourne.

32 *August, 1948—POWER FARMING IN AUSTRALIA AND NEW ZEALAND*

More explanation of the Ferguson System and showing the interdependence of tractor and implement. August 1948.

Normally Ferguson implements were fully mounted on the tractor but these semi-trailing discs were an exception. However they did make use of the three point linkage to adjust the angle of the disc gangs by raising or lowering them. October 1948.

There were massive exports of Ferguson tractors to Australia and New Zealand. May 1948.

Automatic Hitch For New Ferguson Trailer

IMPROVED 3-TON TRAILER! Outstanding features of the new, improved "Ferguson" 3-ton hydraulic tipping trailer are automatic hitching, more traction over soft ground, lower loading level and lower price (£133 ex works, including hitch). You save time by working faster, more efficiently, and where other trailers bog down.

ATTACHED IN 20 SECONDS! Automatic hitching of tractor and trailer is easily made from the driver's seat, mechanical latch makes hitch hold fast. With a loading line of only 2 feet 7 inches, and weighing no more than 16 cwts., the new "Ferguson" tipping trailer is the handiest, most manœuvrable and economical form of farm transport.

TIPPED IN 1 MINUTE! Finger-tip hydraulic control from the driver's seat enables you to tip the trailer a maximum of 40° in less than a minute. Ask your "Ferguson" Dealer for a demonstration on your farm. Let him show you how the "Ferguson" hydraulic tipping trailer saves hauling time. (Non-tipping model £105.10.0 ex works, including automatic hitch).

Farm Better, Farm Faster with FERGUSON

"Ferguson" Tractors are manufactured by The Standard Motor Co. Ltd. for Harry Ferguson Ltd., Coventry

S.22

The improved Ferguson trailer with more commonly seen hitch design. 1949.

The Ferguson Tractor

Pay nothing extra for these exclusive Ferguson features

1. Unique hydraulic mechanism giving finger-tip and automatic control of implements.
2. Ferguson 3-point linkage.
3. Power take-off.
4. Easy wheel-spacing from 48" to 76", in 4" steps; no steering mechanism adjustments needed.
5. Pneumatic tyres.
6. Self starter.
7. Coil Ignition.
8. 4 forward gears giving working speeds from 2½ to 13½ m.p.h. at 1,500 r.p.m.
9. Adjustable drawbar for use with non-Ferguson implements.

Ask your Ferguson Dealer for a demonstration on your farm.

Farm Better, Farm Faster with FERGUSON

"Ferguson" Tractors are manufactured by the Standard Motor Co. Ltd. for Harry Ferguson Ltd., Coventry

S25A

£325 bought you the most advanced tractor. These days it would just about buy a set of tyres and tubes for a Ferguson tractor. 1949.

PETROL OR VAPORISING OIL ?

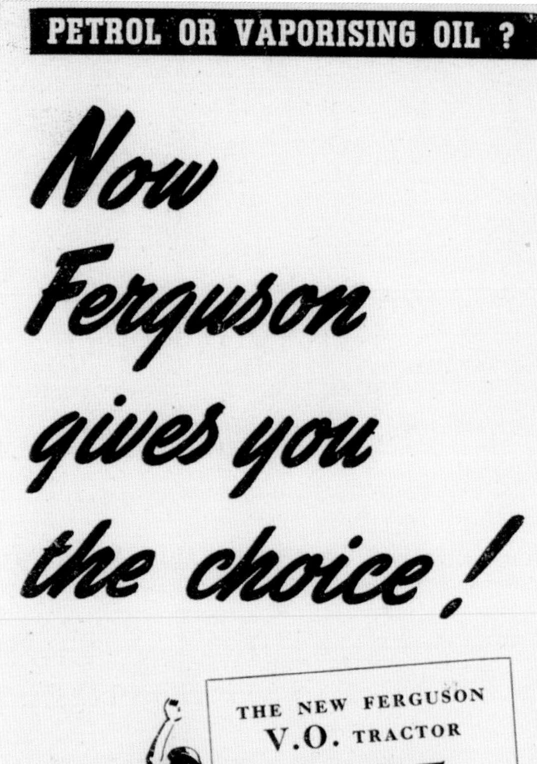

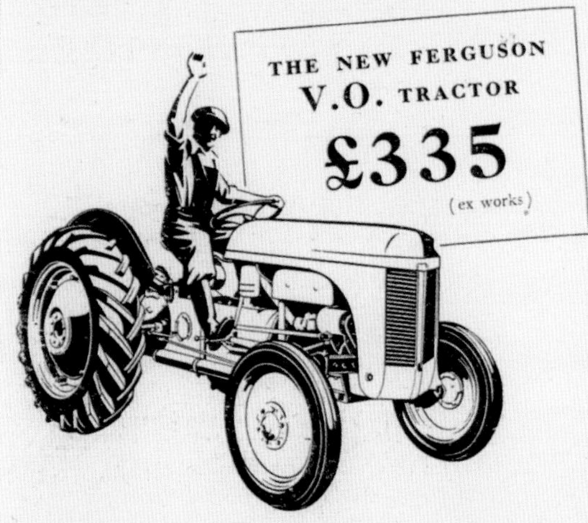

Now Ferguson gives you the choice!

THE NEW FERGUSON
V.O. TRACTOR
£335
(ex works)

Whichever fuel you prefer, vaporising oil or petrol, you can now enjoy the unique advantages of the Ferguson System with its wide range of implements. The new V.O. model is easy to start, its construction is rugged, and extra power is developed at slow and medium engine speeds. Except for the change of engine it is similar in every respect to its £325 petrol equivalent.

Ask your Ferguson Dealer for a demonstration on your farm today! Ask, too, about the Ferguson Pay-as-you-Farm Plan.

Whichever model you choose ....

IT WILL PAY YOU, TOO—TO FARM WITH

Ferguson

S 36A

Ferguson Tractors are manufactured by The Standard Motor Co. Ltd. for Harry Ferguson Ltd., Coventry

Early UK Ferguson tractors were petrol only. When the TVO model was introduced it cost £10 extra, the essential modifications being a twin fuel tank, lower compression engine and a heat shield over the manifold. 1949.

Automatic Hitch for New Ferguson Trailer

ATTACHED IN 20 SECONDS! Outstanding features of the new, improved Ferguson 3-ton trailer are automatic hitching, more traction over soft ground, lower loading line and lower price. Non-tipping model, £99 10s. 0d. ex works; hydraulic tipping model, £127 ex works. (Automatic hitch, £6 ex works.) You save time by working faster, more efficiently and where other trailers bog down. Ask your Ferguson Dealer for a demonstration on your farm and about the Ferguson Pay-as-you-Farm Plan.

Farm Better, Farm Faster with FERGUSON

Ferguson Tractors are manufactured by The Standard Motor Co. Ltd. for Harry Ferguson Ltd., Coventry S22C

Only £6 for a hitch whereby you don't have to leave the tractor seat to hitch up, and then gain all the advantages of weight transfer from the trailer. March 1949.

MANURE SPREADER · MANURE LOADER

Only Ferguson

OFFERS SUCH A FINE RANGE OF IMPLEMENTS—25!

WOOD SAW

FERGUSON PETROL TRACTOR £325. V.O. Model £335
All prices ex works

Farmers' own records prove again and again that *only* Ferguson meets the requirements of *both* large and small farms. Whether you farm 10 acres or 1,000 acres, you can increase production and cut costs with the Ferguson System and any of 25 unique Ferguson implements.

SEE FOR YOURSELF! Ask your Ferguson Dealer to demonstrate the advantages of the Ferguson System on your farm. Ask too about the Ferguson Pay-as-you-Farm Plan.

10" & 12" 2-FURROW PLOUGHS · 10" 3-FURROW PLOUGH · 16" SINGLE FURROW PLOUGH · TILLER · 3-ROW RIDGER · RIGID & SPRING TINE CULTIVATORS · TANDEM DISC HARROW · SPIKE & SPRING TINE HARROWS · POTATO PLANTER · POTATO SPINNER · EARTH SCOOP · TRANSPORT BOX WEEDER HYDRAULIC JACK · STEERAGE HOE · POWER MOWER · SUB-SOILER

3-TON HYDRAULIC TIPPING TRAILER, £130. Non-tipping Model, £99 10s. Automatic Hitch, £6.

IT WILL PAY YOU, TOO—TO FARM WITH Ferguson

Ferguson Tractors are manufactured by The Standard Motor Co. Ltd., for Harry Ferguson Ltd., Coventry. S 38

Now offering 25 implements for use on farms of 10-1000 acres.

Less than three years after production of Ferguson tractors started in UK, it had been adopted in 56 countries – no mean achievement. April 1949.

BRITISH FARM MECHANIZATION April, 1949
"We must give the farmer something which will cut his costs, cut them to pieces—not on a beautiful level field or in a particular crop, but everywhere, under all conditions, all over the world."
—HARRY FERGUSON
Ferguson in action in North Africa—where besides the superior traction and manoeuvrability of the Ferguson System, ease of handling and operating permits the use of unskilled labour.
The famous Ferguson Tractor—prime mover of the Ferguson System.
"... everywhere, under all conditions, all over the world."
Primary objective of every farmer everywhere is to produce more food at less cost, to increase his profitable acreage while cutting his operating costs.
Today, the Ferguson System of complete farm mechanisation is enabling farmers in 56 countries to achieve this objective. You can do the same—regardless of the size of your farm, the nature of its soil or the type of crops you raise. The Ferguson System was specifically designed to meet the needs of farmers everywhere, under all conditions.
You can work a Ferguson tractor and implement in small fields and irregularly shaped plots. You can farm up and across steep hillsides accurately and with ease. You can farm hard or rocky soil without fear of damage to tractor
2-Furrow Plough
Tiller
Disc Harrow
Sub-Soiler
Spring Tine Cultivator
Mower
FARM BETTER, FARM
April, 1949 BRITISH FARM MECHANIZATION
Ferguson tractors for farmers in the eastern hemisphere are produced by The Standard Motor Co. Ltd., Coventry, England, and Ferguson implements by foremost British manufacturers. Ferguson tractors and implements for farmers in the western hemisphere are supplied by Harry Ferguson, Inc., Detroit, Michigan, U.S.A.
The modern 80-acre Banner Lane factory of The Standard Motor Co. Ltd., Coventry, England.
Ferguson System cuts cost of farming in 56 countries
or implement. You can work where other implements won't. You can farm from the driver's seat.
Prime mover of the Ferguson System is the lightweight Ferguson tractor that automatically converts itself into a heavy tractor for heavy jobs. This means you get more traction without excess built-in weight that packs the soil. Simple 3-point linkage in one minute and hydraulic control—both manual and automatic—co-ordinate the action of tractor and implement to give maximum versatility, manoeuvrability and ease of handling. The result is easier farming, better farming, more profitable farming.
Illustrated below are some of the Ferguson implements designed to meet a multitude of farming jobs.
FERGUSON SYSTEM

Versatile Ferguson Makes Row Crop Work Easy
WHEEL SPACE VARIATIONS! You can do all types of row crop work with the Ferguson. Rear wheels can be spaced from 48″ to 76″ (front 48″ to 80″) or offset as desired, without change of steering mechanism. In action above is the Spring Tine Cultivator. Automatic hydraulic control maintains working depth. It's part of the Ferguson System of complete farm mechanisation. Ask your Ferguson Dealer for a demonstration on your farm and about the Ferguson Pay-as-you-Farm Plan.
FARM BETTER, FARM FASTER WITH FERGUSON
FERGUSON SYSTEM
FERGUSON TRACTORS ARE MANUFACTURED BY THE STANDARD MOTOR CO. LTD. FOR HARRY FERGUSON LTD., COVENTRY S.181

What could be easier? Not only the track settings permitted by the Ferguson wheel and axle designs, but also the ease of adjustment of the spring tines to suit any minor variations in row width or stage of plant growth.

MOW without stopping, backing or circling
THIS TRACTOR-MOUNTED POWER MOWER is raised or lowered by finger-tip control from the driver's seat. You can cut a 5-ft. swath at speeds up to 4½ m.p.h. more cheaply than ever before, and cover from 20 to 30 acres per day. Location of cutter bar permits mowing of square corners without stopping, backing or circling. Safety spring release protects cutter bar from hidden obstructions. Ask your Ferguson Dealer for a demonstration on your farm and about the Ferguson Pay-as-you-Farm Plan.
10-acre nurseryman, Dick Wheeler, of Mytchett Nurseries near Farnborough, Hants., says—
"It will pay you, too—to farm with Ferguson"
FERGUSON SYSTEM
Ferguson tractors are manufactured by The Standard Motor Co. Ltd., for Harry Ferguson Ltd., Coventry
532A

Anyone who has operated a Ferguson mower will know that it was a reliable and efficient tool but a dreadful implement to hitch and unhitch. 1949

32 BRITISH FARM MECHANIZATION August, 1949 | August, 1949 BRITISH FARM MECHANIZATION 33

One t[illegible]tor, one system, one agricultural method for farmers everywhere

"Until now, no one has come along with machinery which will do, on all the farms of the world, everything which animals now do—and do it at a fraction of the cost."

—HARRY FERGUSON

Never designed for farm[illegible], but adapted to it out of necessity, the power animal has met its maste[illegible]day, in 56 countries, farmers are proving that the Ferguson System of [illegible] farm mechanisation can solve the problem of the small farm, as well [illegible] large, all over the world.

The Ferguson System [illegible] of two parts: the Ferguson tractor and any Ferguson implement. Co-ordinated by 3-point linkage and both finger-tip and automatic hydraulic control, a Ferguson tractor-implement unit gives farmers their first real chance to farm more efficiently, profitably and with less back-ache. They can produce more food at less cost from every available acre—however few acres they farm.

Australia

Australian farmers appreciate the sturdy construction of Ferguson tractors and implements and the versatility of the Ferguson System.

Britain

British farmers run their Fergusons the year around under all conditions. Better, more profitable farming becomes automatic—and so much easier.

India

Unskilled Indian labour have no difficulty operating Ferguson equipment. Simplicity of design reduces maintenance to a minimum.

Switzerland

More traction and stability enable Fergusons to do yeoman duty in Switzerland — where uphill and sidehill farming is the general rule.

France

No small part of the Ferguson's success in French vineyards is due to the ease with which it can be manoeuvred in restricted areas.

South America

Easy attachment of Ferguson implements enables farmers in South America to change quickly from one Ferguson implement to another.

Ferguson tractors and implement for world-wide distribution are produced by the most modern methods and equipment. Tractors for the eastern hemisphere are produced in the 90-acre factory of the Standard Motor Co. Ltd., Coventry, England, and implements by foremost British manufacturers. Farmers in the western hemisphere are supplied from the great new 72-acre plant of Harry Ferguson, Inc., Detroit, Mich., U.S.A.

FARM BETTER, FARM FASTER WITH FERGUSON

FERGUSON SYSTEM

Ferguson advertising relentlessly pressing home the message of the Ferguson's suitability for all the world's farmers. June 1949.

Going for colour to impress the world – and succeeding! August 1949.

THE FERGUSON FARMER IS BACKED BY A MIGHTY BRITISH ORGANIZATION
THE BANNER LANE WORKS OF STANDARD MOTOR CO. LTD
THE FERGUSON SYSTEM
500 UNITS PER DAY are rolling off the line at the Ferguson System factories in Britain. The aerial view above shows the vast Banner Lane works of the Manufacturers at Coventry, England. Here over 1,000,000 square feet of factory space is devoted exclusively to the manufacture of Ferguson System Tractors. Similar factories are making Ferguson implements. The British Farm Equipment Company proudly presents the quality and service of the Ferguson System, backed by the prestige of BRITISH MANUFACTURE. The Ferguson System of farming has been developed through 30 years' experience with all types of agriculture. Already there are 90 approved Ferguson Dealers in N.S.W., all trained to service the tractor and implements, and able to supply spare parts through the years ahead.
This labour-saving, economic equipment is being shipped out in big volume and is AVAILABLE HERE NOW!
F11/248
BRITISH FARM EQUIPMENT CO.
(DIVISION OF STANDARD CARS LTD.)
83-89 FLINDERS STREET, SYDNEY
and 568-576 Elizabeth Street, Melbourne.
48
May, 1949—POWER FARMING IN AUSTRALIA AND NEW ZEALAND

Southern hemisphere advertising of the UK Banner Lane plant. May 1949.

Introducing the new 85mm Ferguson tractor engine down under. August 1951.

Amazing Success of FERGUSON SYSTEM gives you the benefit of DRASTIC PRICE REDUCTIONS!

In less than a year the British Ferguson System of farming has *proved itself* to thousands of countrymen throughout the State. The great Ferguson story is one of outstanding economy and uncanny efficiency on all types of farm work in all kinds of soil.

This unqualified success allows us to *drastically reduce the price* of the Ferguson tractor and implements when practically every other farming requirement is costing you more. In these days of steeply rising costs, Ferguson offers you better farming and greater efficiency, *at a greatly reduced capital and running cost*, thus contributing, in no small measure, to the stabilisation of the National economy.

See your local Ferguson dealer now for the exact savings on various items. These range from £15 on the tractor itself to £21 on Cultivators. *Start your New Year well with a complete farming system at a substantial saving in outlay!*

Ferguson offers you IMMEDIATE DELIVERY. *Big shipments* are arriving constantly.

FARM BETTER, FARM FASTER WITH FERGUSON

F18/249

THE FERGUSON SYSTEM

BRITISH FARM EQUIPMENT CO.
(Division of Standard Cars Ltd.)
83-89 FLINDERS STREET, SYDNEY
568-578 ELIZABETH ST., MELBOURNE, C.1

22 March, 1949—POWER FARMING IN AUSTRALIA AND NEW ZEALAND

The Ferguson System of farming going well in Australia and New Zealand. March 1949.

There was always a shortage of manpower on farms down under – Ferguson to the rescue! October 1949.

Novel advertising – Ferguson for rabbit control! July 1949.

There was wide adoption of the Ferguson tractor in New South Wales in 1949 across the farming spectrum. September 1949.

A splendid colour advertisement emphasising Ferguson's world wide market, world wide application of his mechanisation system and his global visions of fighting poverty and hunger. 1949.

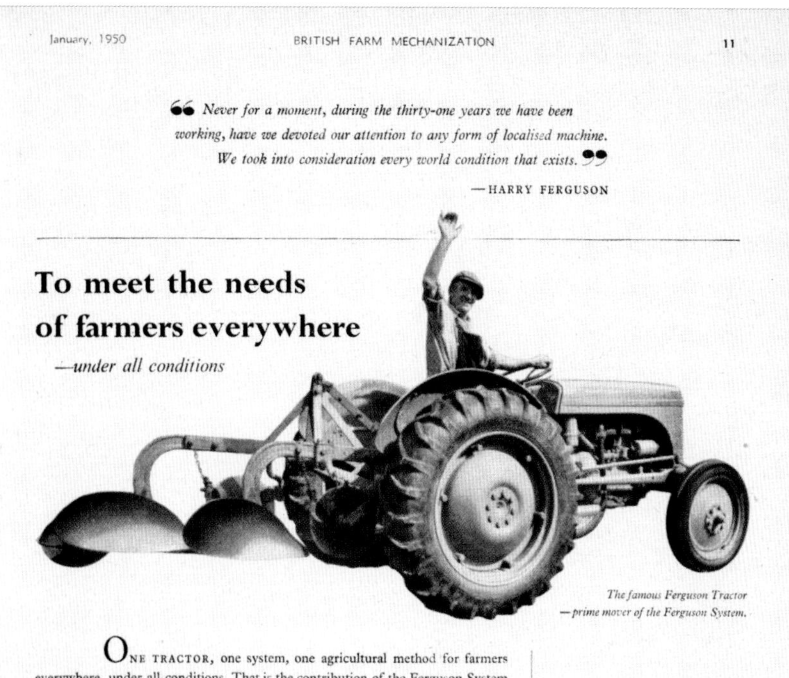

January, 1950
BRITISH FARM MECHANIZATION
11
"Never for a moment, during the thirty-one years we have been working, have we devoted our attention to any form of localised machine. We took into consideration every world condition that exists."
— HARRY FERGUSON
To meet the needs of farmers everywhere
—under all conditions
The famous Ferguson Tractor
—prime mover of the Ferguson System.

ONE TRACTOR, one system, one agricultural method for farmers everywhere, under all conditions. That is the contribution of the Ferguson System of *complete* farm mechanisation. It is a contribution that gives farmers their first real chance to farm as they have always wanted to farm.

What is the Ferguson System?

The Ferguson System consists of two parts: the Ferguson tractor and a Ferguson implement. Simple 3-point linkage and hydraulic control — both manual and automatic — co-ordinate the action of the tractor and implement. They work together as a single unit to provide these six advantages:

1. *Traction without excess built-in weight.*
2. *Penetration without weight.*
3. *Hydraulic depth control of the implement.*
4. *Automatic protection from hidden obstructions.*
5. *Easy attachment of implements in one minute.*
6. *Tractor front end stays down — even uphill.*

The sum total of these advantages is easier farming, better farming, more profitable farming. Today, farmers in 56 countries are proving that the Ferguson System enables them to produce more food at less cost — regardless of the size of their holdings, the nature of the soil or the type of crops grown.

Ferguson tractors for farmers in the eastern hemisphere are produced by The Standard Motor Co. Ltd., Coventry, England, and Ferguson implements by foremost British manufacturers. Ferguson tractors and implements for farmers in the western hemisphere are supplied by Harry Ferguson, Inc., Detroit, Michigan, U.S.A.

N.4

IT WILL PAY YOU, TOO—TO FARM WITH·FERGUSON

Harry Ferguson looks back 31 years to when he started with his mechanical lift and weight transfer design of ploughs. Not a man to look backwards, he must at this stage have been taking some satisfaction from the success of his Ferguson System all over the world. This stemmed from his initial appreciation of the inadequacy and inefficiency of the traditional single point hitching of implements to tractors. January 1950.

January-February, 1950 FARM MECHANIZATION

Only FERGUSON

MEETS REQUIREMENTS OF BOTH LARGE AND SMALL FARMS

FOR LARGE FARMS

"Incorporating entirely new methods of implement linkage and control, the Ferguson System has revolutionised power farming, and *all large scale operators* are advised to study this system in the light of their individual needs."

—Mr. A. H. Hoare, Ministry of Agriculture (in his recent book "Vegetable Crops for Market").

The Ferguson Manure Loader and Spreader

Saves time, cuts labour costs, because *one man* can haul and spread 25 loads a day without fatigue.

FOR SMALL FARMS

"The Ferguson System is available equally to the *smaller farmer* and *market gardener*. The capital outlay is within their reach, and they are able to derive the same benefits as the large scale grower.

—Mr. L. M. Marshall (the well-known contributor to farming journals).

The Ferguson Spring Tine Cultivator

Alignment of tractor front wheels and cultivator is maintained by automatic steering fin. The light, manoeuvrable tractor can be easily adjusted to required row widths—keeps unproductive headland to a minimum.

The Ferguson System has been designed to help farmers of all kinds *produce more food at less cost. Whether you farm 10 acres or 1,000 on any kind of soil . . .*

Ask your Ferguson Dealer for a demonstration on your farm and about the Ferguson Pay-as-you-Farm Plan.

IT WILL PAY YOU, TOO—TO FARM WITH Ferguson

FERGUSON TRACTORS ARE MANUFACTURED BY THE STANDARD MOTOR CO. LTD., FOR HARRY FERGUSON LTD., COVENTRY. S.39

Praise from both the small and large farmer. January 1950.

8
BRITISH FARM MECHANIZATION
February, 1950
would you dig a
well with a spoon?
Dig a well with a spoon! Certainly not, you're bound to agree—not unless you hadn't any other choice.
On the other hand, this proposition is not as ridiculous as it seems when you remember that the methods of production today on the majority of the world's farms are equally antiquated and inefficient. Actually the farmer's equipment is substantially as it was hundreds of years ago.
Why? Because until the conception and practical application of the Ferguson System of *complete* farm mechanisation, no one had come along with machinery which would do, on all the farms of the world, everything which animals could do—*and do it at a fraction of the cost*. Today the Ferguson System—and the wide range of implements specifically designed for use with it—is enabling farmers in fifty-six countries to retire the power animal in favour of this more practical, efficient and economical agricultural method. In their new-found ability to produce more food at less cost, these farmers are making a substantial contribution to the present and future prosperity of their countries.
FERGUSON SYSTEM
THE FERGUSON SYSTEM MEANS AGRICULTURAL PROSPERITY
Ferguson tractors are manufactured by The Standard Motor Co. Ltd., for Harry Ferguson Ltd., Coventry, England.
N.10

Wherever you farmed in the world the message was the same – replace animal power with a Ferguson. February 1950.

MAR. 1, 1950 FARM IMPLEMENT AND MACHINERY REVIEW 1469

"Best investment I ever made"

— says Mr. D. G. Rickards of his Ferguson tractor and implements.

Mr. Rickards, his son, and a land-girl, farm 124 acres at Tarbay Farm, Oakley Green, Windsor. "It was always a race against time until I bought the Ferguson," says Mr. Rickards.

Out of 'pony' class

"I purchased the Ferguson with the idea that it would do the 'pony' work instead of the big tractor, but after two or three days it soon came out of the 'pony' class and was used for more and more general work. Mowing, muck lifting, silage carting and general hauling are all done by Ferguson now, and it has never let us down. Its ease of handling, besides saving time, makes for increased production.

Dick Rickards and the Ferguson make light work of uphill ploughing at Tarbay Farm.

Ploughing more advanced.

"Most important of all, though, is the ploughing. We have to contend here with good solid clay, and my son doubted whether the Ferguson could tackle this kind of work—especially as his bigger tractor had refused. Still, we tried, and thanks to the Ferguson tractor and plough last autumn's ploughing was more advanced than ever before. With the Ferguson we can cultivate odd corners and marginal land that has never been ploughed before. It's certainly the best investment I ever made for the farm."

Ask your Ferguson Dealer for a demonstration on your farm and about the Ferguson Pay-as-you-Farm Plan.

ON ANY SIZE FARM, IT WILL PAY YOU, TOO—TO

FARM WITH FERGUSON

S.40

Ferguson tractors are manufactured by The Standard Motor Co. Ltd., for Harry Ferguson Ltd., Coventry.

Not just a pony, but a tractor which can handle clay that larger tractors refused. March 1950.

The horse is glad of Ferguson-induced retirement! May 1950.

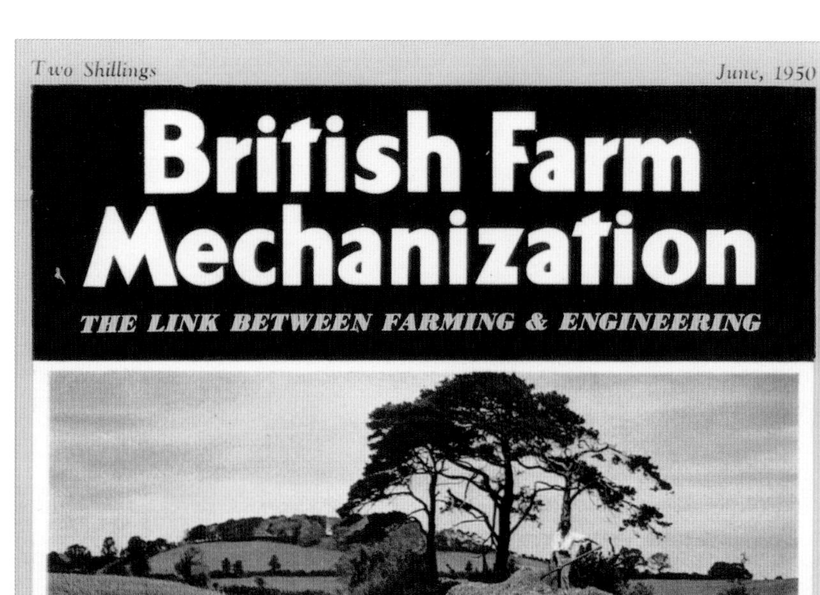

One missing aspect of Ferguson versatility in its implement range was a green crop loader to enable the pitchfork to be retired with the horses! June 1950.

Recommendations of the value of the Ferguson system just kept coming in. June 1950.

JUNE 1, 1950 FARM IMPLEMENT AND MACHINERY REVIEW 203

All over Britain

THEY'RE TALKING THIS WAY ABOUT FERGUSON

"Biggest thing in years"

"'The hydraulic lift perfected by Ferguson has been the biggest thing we have had these last few years,' Mr. Aspinwall declared . . . Mr. Basil Hayes . . . claimed that the Ferguson beat anything America was putting out for its size . . . ."

(Extracts from a discussion reported by "The Preston Guardian," 7 Jan., 1950.)

"Best investment I ever made"

"It was always a race against time until I bought the Ferguson. Mowing, muck lifting, silage carting and general hauling are all done by Ferguson now, and it has never let me down. It's certainly the best investment I ever made."

Mr. D. G. Rickards, Tarbay Farm, Oakley Green, Windsor.

"The best thing I ever bought in my life"

"The time we've saved since we've had the Ferguson is wonderful. We use it for everything — ploughing, mowing, hauling—and it's the best thing I ever bought in my life."

Mr. G. S. Smith, East End Farm, Nailsea, Nr. Bristol.

"Pulls twice as much uphill"

"We believed we should get convenience of unloading and manœuvrability in the Ferguson tipping trailer . . We were satisfied, but our surprise has come from the fact that regularly we are taking four-ton loads where before we had to be content with two-ton loads . . . We cannot explain why a 23 b.h.p. pulls twice as much as a 27 b.h.p. tractor, but the reduction in operating costs is such that we feel the facts should be known . . . ."

(Extract from a letter in "The Farming News" from Vitagrass Farms Ltd., Cowans Bridge, Carnforth.)

"Saves labour and money"

"It saves us an enormous amount of labour and money. Spraying used to take us a week, now we do it in half a day."

Mr. Fred Wheeler, Mytchett Nurseries, Nr. Farnborough, Hants.

"Does three men's work"

"I carry at least three head of stock more since I had the Ferguson. And I can now do as much work as three men with horses—more easily."

Mr. J. McLean, Drumballyhue, Rock, Dungannon, N. Ireland.

Wherever you find a Ferguson farmer you'll hear the same story. Ferguson helps you produce more food at less cost from every available acre. Ask your Ferguson Dealer for a demonstration on your farm and about the Ferguson Pay-as-you-Farm Plan.

ON ANY SIZE FARM—IT WILL PAY YOU, TOO—TO

FARM WITH FERGUSON

Plough for plenty
and do a
sterling job

Ferguson tractors are manufactured by The Standard Motor Co. Ltd., for Harry Ferguson Ltd., Coventry

AUG. 1, 1950 FARM IMPLEMENT AND MACHINERY REVIEW 553

Economic Ferguson Farming

ON 12 ACRES

Mr. C. S. Marshall, of Lower Acton Farm, Stourport-on-Severn, Worcester, farms 12 acres plus a small acreage of grass taken separately for cattle grazing. Until three years ago, arable work was done by horse after ploughing by hired tractor. Now the Ferguson does everything. More muck is got on to the land. More crops can be grown. More work is done more quickly—and it's all done by Mr. Marshall himself, without extra labour costs, and without drudgery. In addition, as Mr. Marshall points out, the Ferguson will last for years on a holding of this size. It will be doing good work after paying for itself over and over again.

Mr. Marshall finds his Ferguson wonderfully easy to handle in confined spaces. He even uses it to plough his kitchen garden within a few feet of his back-door.

Mr. Marshall ploughing between the trees in his orchard. The Ferguson is so manoeuvrable that not an inch of ground in his orchard or garden remains uncultivated.

The Ferguson is useful for innumerable odd jobs, for hauling feed to cattle, acting as a power unit for static machinery such as a wood saw, and even, as here for moving chicken arks to a fresh site.

Loosening the top soil round bean plants with the Ferguson spring tine cultivator. The tractor is easily adjusted to row widths. Accurate working depth is maintained hydraulically.

Ask your Ferguson Dealer for a demonstration on your farm and about the Ferguson Pay-as-you-Farm Plan.

FERGUSON SYSTEM

ON ANY SIZE FARM—IT WILL PAY YOU, TOO—TO

Plough for plenty
and do a
sterling job

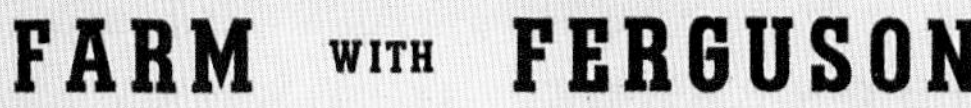
FARM WITH FERGUSON

Ferguson tractors are manufactured by The Standard Motor Co. Ltd. for Harry Ferguson Ltd. Coventry.

S.62

A £325 tractor for a 12 acre farm meant an investment of £2.10s an acre if the tractor is given a life of 10 years. August 1950.

8 BRITISH FARM MECHANIZATION July, 1950

THE *Ferguson* SYSTEM RELIES ON THESE FAMOUS FIRMS

In just three years, Ferguson tractors and implements made in Britain have proved their ability to produce more food at less cost in almost every part of the world. They are saving time and labour — cutting production costs to the bone — on farms of all types and sizes.

One reason for this success is that the Ferguson System of *complete* farm mechanisation is designed to meet every known farming condition. But this theory can only work if the machinery produced is of sufficiently high standard. Ferguson implements, therefore, are precision-built by British manufacturers from the finest materials available. They are built to work more efficiently and last longer wherever the farmer might need them — not merely on level fields — but in bogs and on hills, and on rocky or broken ground. It is because these implements have been manufactured only by famous British firms, embodying the finest traditions of craftsmanship, that the success of the Ferguson system has been possible.

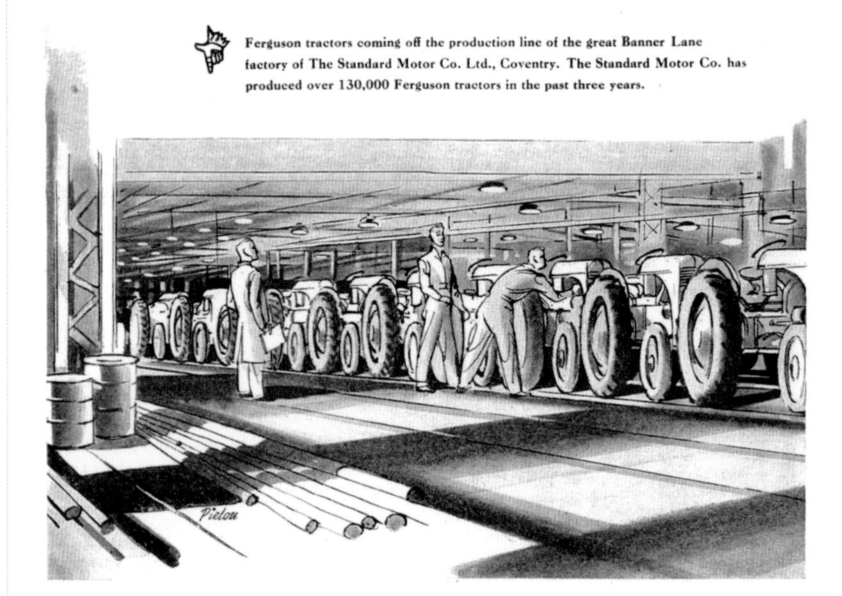

Ferguson tractors coming off the production line of the great Banner Lane factory of The Standard Motor Co. Ltd., Coventry. The Standard Motor Co. has produced over 130,000 Ferguson tractors in the past three years.
Pielou

July, 1950 BRITISH FARM MECHANIZATION 9

Ferguson made only tractors. Here is a rare list of manufacturers who made the implements to Ferguson quality specifications for subsequent badging by Ferguson. July 1950.

12
BRITISH FARM MECHANIZATION
September, 1950
"I'd be lost without FERGUSON"
. . . says 465 acre farmer
Mr. J. G. Fisher farms 465 acres — mostly arable — at Gunnerby Farm, near Hatcliffe, Lincoln. A great deal of the work is done by his two Ferguson tractors with their various implements. The Fergusons work 8 hours a day every day. They carry 3-ton loads for 8 hours daily on 3 gallons of petrol. Last year they drilled 70 tons of fertiliser at the rate of 15 acres a day.
The Ferguson's adaptability, pulling power, and ease of handling are the points that impress Mr. Fisher. "I would not like to tackle such a job without the Fergusons," he says. "In fact I'd be lost without them!" Mr. Fisher's father, who farms 615 acres nearby, is also a confirmed Ferguson user.
Mr. Fisher's experience confirms that of many other farmers — that the Ferguson System is as valuable on large farms as it is on smaller ones. Ferguson tractors and implements do heavy jobs cheaply and well. They are also sufficiently handy and economical to do the hundred-and-one odd jobs that abound on big farms — faster and better. One man with a Ferguson can do all the carrying, shifting, cleaning, lifting, muck spreading, ditch digging, etc. that would ordinarily occupy 3 or 4 men. In large farms this means not only a saving of labour, but also a more economical distribution of labour. It means more food at less cost on every available acre.
Mr. Fisher with his Ferguson Disc Plough. The Wold soil varies from heavy clay to light chalk, and the Ferguson does a consistently good job under all conditions.
The Ferguson pulling three Cambridge crimp rollers up one of the steep Gunnerby fields. Non-Ferguson implements are easily attached by means of the drawbar.
September, 1950
BRITISH FARM MECHANIZATION
13
TWO NEW FERGUSON IMPLEMENTS
The Ferguson Disc Plough. With 2 or 3 furrows. Special tubular lightweight frame, and 3-point linkage produces unrivalled ease of handling. Hydraulically controlled from driver's seat. Ideal for spring and cross-ploughing, and for especially hard, gritty or sticky land.
The Ferguson Post Hole Digger. Power driven. Operated hydraulically from the driver's seat. Drills holes of 9" or 12" diameters and 3' deep in less than 30 seconds. Replaceable screw points and cutting blades. Works equally well on slopes and other difficult positions.
Plough for plenty and do a sterling job
Ask your Ferguson Dealer for a demonstration on your farm, and about the Ferguson Pay-as-you-Farm Plan.
Ferguson tractors are manufactured by The Standard Motor Co. Ltd., for Harry Ferguson Ltd., Coventry.
ON ANY SIZE FARM, IT WILL PAY YOU, TOO – TO FARM WITH FERGUSON

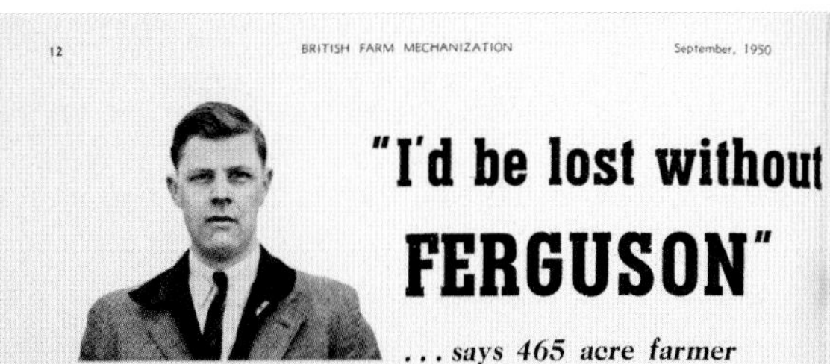

Never a common sight on British farms but here a large farmer extols the virtues of the Ferguson disc plough which was more widely sold overseas in drier farming areas. September 1950.

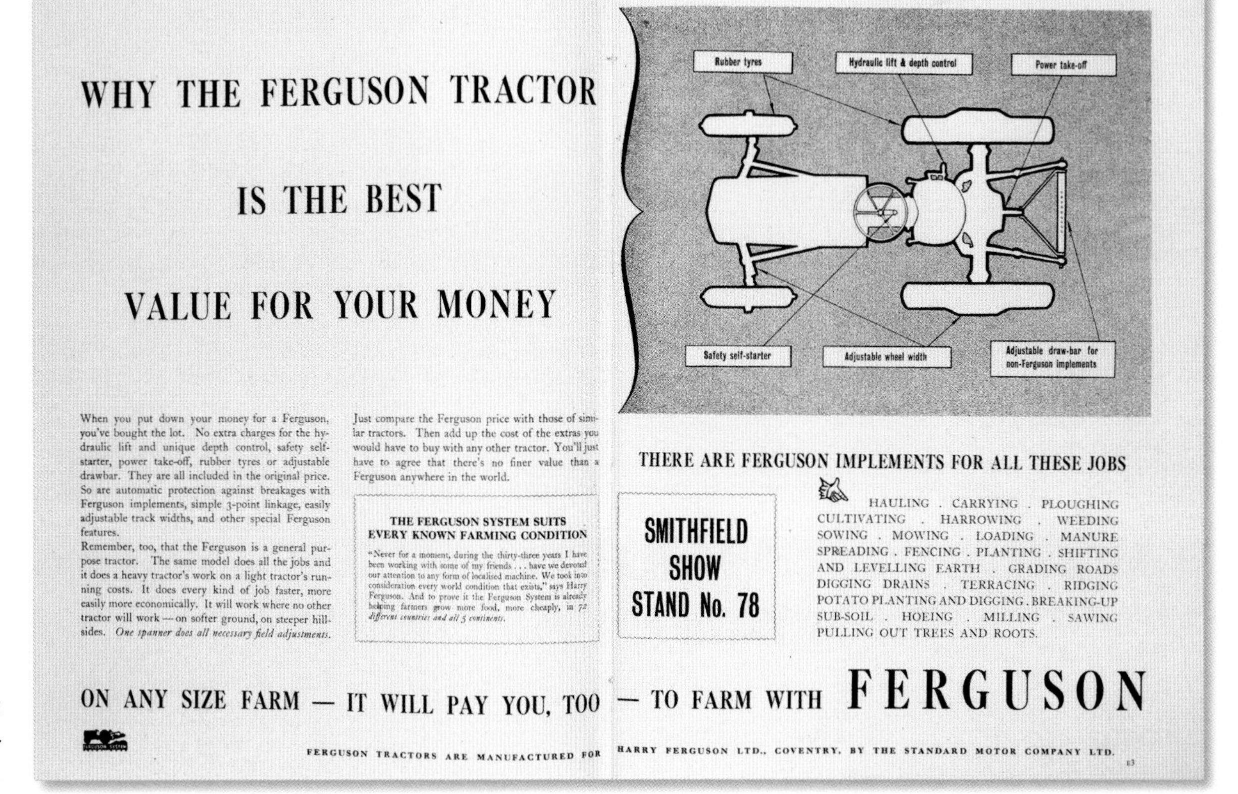

34
FARM MECHANIZATION
November-December, 1950
WHY THE FERGUSON TRACTOR
IS THE BEST
VALUE FOR YOUR MONEY
When you put down your money for a Ferguson, you've bought the lot. No extra charges for the hydraulic lift and unique depth control, safety self-starter, power take-off, rubber tyres or adjustable drawbar. They are all included in the original price. So are automatic protection against breakages with Ferguson implements, simple 3-point linkage, easily adjustable track widths, and other special Ferguson features.
Remember, too, that the Ferguson is a general purpose tractor. The same model does all the jobs and it does a heavy tractor's work on a light tractor's running costs. It does every kind of job faster, more easily more economically. It will work where no other tractor will work — on softer ground, on steeper hillsides. One spanner does all necessary field adjustments.
Just compare the Ferguson price with those of similar tractors. Then add up the cost of the extras you would have to buy with any other tractor. You'll just have to agree that there's no finer value than a Ferguson anywhere in the world.
THE FERGUSON SYSTEM SUITS EVERY KNOWN FARMING CONDITION
"Never for a moment, during the thirty-three years I have been working with some of my friends . . . have we devoted our attention to any form of localised machine. We took into consideration every world condition that exists," says Harry Ferguson. And to prove it the Ferguson System is already helping farmers grow more food, more cheaply, in 72 different countries and all 5 continents.
November-December, 1950
FARM MECHANIZATION
35
Rubber tyres
Hydraulic lift & depth control
Power take-off
Safety self-starter
Adjustable wheel width
Adjustable draw-bar for non-Ferguson implements
THERE ARE FERGUSON IMPLEMENTS FOR ALL THESE JOBS
HAULING . CARRYING . PLOUGHING
CULTIVATING . HARROWING . WEEDING
SOWING . MOWING . LOADING . MANURE
SPREADING . FENCING . PLANTING . SHIFTING
AND LEVELLING EARTH . GRADING ROADS
DIGGING DRAINS . TERRACING . RIDGING
POTATO PLANTING AND DIGGING . BREAKING-UP
SUB-SOIL . HOEING . MILLING . SAWING
PULLING OUT TREES AND ROOTS.
SMITHFIELD SHOW STAND No. 78
ON ANY SIZE FARM — IT WILL PAY YOU, TOO — TO FARM WITH FERGUSON
FERGUSON SYSTEM
FERGUSON TRACTORS ARE MANUFACTURED FOR HARRY FERGUSON LTD., COVENTRY, BY THE STANDARD MOTOR COMPANY LTD.

Now it's value for money in five continents and 72 countries, and only one spanner needed for all field adjustments. November 1950.

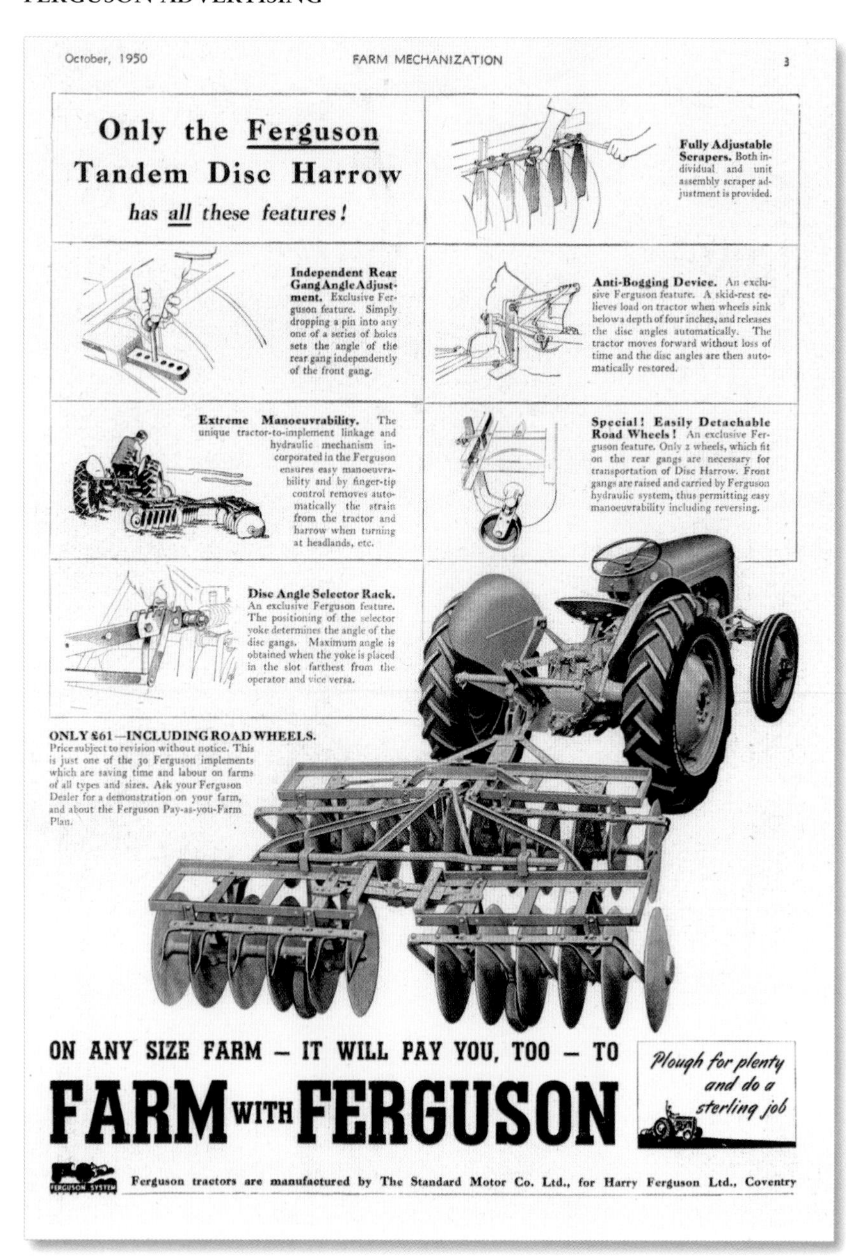

Quite rare today are Ferguson trailed disc harrows which were to be replaced in time with the Ferguson mounted type. This harrow is a good example of Ferguson hydraulic power being used for on the move adjustment of the disc gangs' angles of work. Does anyone know of a set of surviving road wheels for the implement? October 1950.

Ferguson goes poetic for Christmas. December 1950.

DEC. 1, 1950 FARM IMPLEMENT AND MACHINERY REVIEW 1217

150,000
FERGUSON
TRACTORS SOLD
SINCE 1946

IT WILL PAY YOU, TOO – TO

FARM WITH FERGUSON

FERGUSON TRACTORS ARE
MANUFACTURED BY THE STANDARD MOTOR CO., LTD.,
FOR HARRY FERGUSON LTD., COVENTRY

Still less than a third of the final number of Ferguson tractors that were to be produced in the UK by the end of 1956 – Ferguson carries on replacing animal power. December 1950.

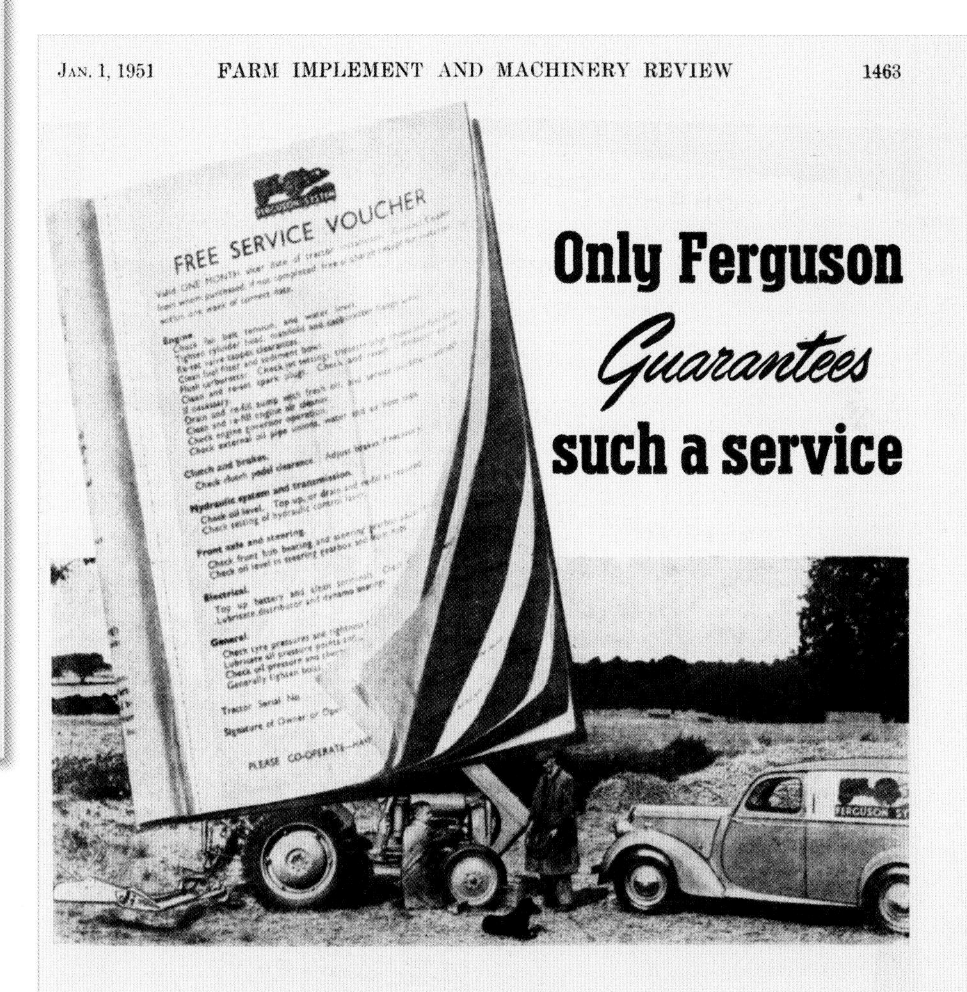

JAN. 1, 1951 FARM IMPLEMENT AND MACHINERY REVIEW 1463
FREE SERVICE VOUCHER
Only Ferguson
Guarantees
such a service

Free Service—Voucher Scheme

With 4 simple service vouchers in your Ferguson tractor Instruction Book, you are guaranteed free service to your tractor during the warranty period of 6 months. Simply by presenting the vouchers at the first, second, fourth and sixth month to any Ferguson Dealer, the work detailed on each voucher will be carried out at the periods stated and at your convenience, without cost to yourself other than expendable parts.

It's all part of Ferguson's policy of giving farmers the finest possible after-sales service and ensures that your Ferguson tractor starts its life right.

ON ANY SIZE FARM IT WILL PAY YOU, TOO - TO

FARM WITH FERGUSON

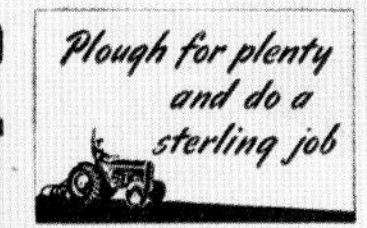

FERGUSON TRACTORS ARE MANUFACTURED FOR HARRY FERGUSON LTD., COVENTRY, BY THE STANDARD MOTOR COMPANY LTD. S.57

Six months warranty and four service vouchers started your life with a new Ferguson tractor. January 1951.

June 1, 1951 FARM IMPLEMENT AND MACHINERY REVIEW 187
TIDY UP YOUR FARM
and turn scrap into cash!
Here is a new Ferguson message for farmers
Britain needs more steel. She needs it for the armed forces and agriculture, for defence, and for export. And every ton of scrap metal recovered means an extra ton of new steel. Every bit of scrap — every piece of obsolete and inefficient machinery — that you can clear off your farm, puts more steel into production and money into your pocket for up-to-date equipment.
The more steel produced, the more steel will be available for manufacture . . . including the manufacture of Ferguson tractors. It's only common sense, therefore, for Harry Ferguson Ltd., to help salvage scrap metal. Your local Ferguson Dealer will be pleased to put you in touch with your nearest Scrap Metal Merchant, or with the Joint District Scrap Committee in your area. You'll be able to tidy up your farm, get a good price for the scrap, and help Britain into the bargain.
Raw materials are getting scarcer. Although iron and steel production broke new records in 1950, the increase can only be maintained if the maximum amount of scrap metal from home sources is sent back to the furnaces.
You provide the scrap and we'll provide the tools
Ferguson tractors are manufactured for HARRY FERGUSON LTD., Coventry, by The Standard Motor Co., Ltd.

Ferguson appeals to farmers to turn in their scrap metal. Britain was still facing war period induced shortages of steel even into the 1950s. June 1951.

How the Ferguson System fights hunger & poverty

The world's population is increasing faster than food production. And dear food—scarce food—is causing high living costs and world unrest. Slow, laborious farming with power animals or inefficient machinery can never solve this food problem. But a solution has been found—with the Ferguson System of *complete* farm mechanisation.

In 4 years Harry Ferguson Ltd., Coventry, have sold 150,000 tractors and the implements to work with them. These represent not merely new machinery but a new farming system. A system that is working successfully in 76 different countries.

The Ferguson System combines all the advantages of light and heavy machinery. It costs less to buy, less to run, and less to maintain. It enables old men and women, boys and girls to do a strong man's work — faster, better, more cheaply than ever before. This system has already produced up to *ten times more food* in some areas. It is helping farmers *everywhere* produce more food at less cost from every available acre.

GROW MORE FOOD — MORE CHEAPLY — WITH Ferguson

Ferguson tractors are manufactured for Harry Ferguson Ltd., Coventry, by The Standard Motor Company Ltd.

This colour advertisement was placed in journalists' material produced for the Festival of Britain. One of Harry Ferguson's great hopes was that he could help reduce hunger and poverty. The Festival was an excellent venue to expound upon this theme. 1951.

Two Shillings *June, 1951*

Farm Mechanization

THE LINK BETWEEN FARMING & ENGINEERING

The Ferguson tractor-mounted power mower cuts a 5-foot swathe at speeds up to 3½ m.p.h. under all conditions. Location of cutter bar just behind rear tractor wheels, plus independent rear wheel brakes allows quick, easy turns. Raised and lowered by finger-tip hydraulic control, the mower is also protected against damage from hidden obstacles by a safety spring release. The higher-than-average knife speed deals efficiently with short grass for silage and drying.

On any size farm, it will pay you, too — to

FARM WITH FERGUSON

Ferguson tractors are manufactured for HARRY FERGUSON LTD., COVENTRY, by The Standard Motor Company Limited

Some artistic licence has been expressed on this field of grass – it seems to be dry before it has been cut! The faster than average Ferguson knife speed is vibrating the small print but not the main message – Farm With Ferguson. June 1951.

IN 60 DIFFERENT COUNTRIES THROUGHOUT THE WORLD

THE NEW, MORE POWERFUL

Ferguson PETROL TRACTOR

IS WELCOMED

(an increase of 5½ h.p. . . . AT NO EXTRA COST!)

Another step in the progress of the Ferguson System. Another chapter in the story of faster and better farming. The new, more powerful Ferguson Tractor comes to you with these specifications:—

ENGINE:

TE-A-20 petrol type, four-cylinders (wet sleeves). Bore, 85 mm. Stroke, 92 mm. Piston displacement, 127.4 cu. in. (2,088 c.c. swept volume). Compression ratio, 6.1. Firing order, 1, 3, 4, 2.

Horse Power: 27 B.H.P. at 2,000 r.p.m.

Fuel System: Welded steel tank below hood.

Capacity, eight gallons. Reserve of one gallon, held and operated by two-way valve.

MODIFICATIONS INTRODUCED AFFECT THESE COMPONENTS ONLY:

• Cylinder bore and compression ratio • Cylinder head and water pump • Cylinder block, sleeves, pistons • Camshaft and valve gear • Fuel and ignition systems • Radiator cowl and thermostat body • Clutch.

BRITISH FARM EQUIPMENT CO.
(Division of Standard Cars, Ltd.)
83-97 Flinders Street, Sydney. 'Phones: FA 2083, FA 4183.
568-578 Elizabeth Street, Melbourne. 'Phone: FJ 2154.

BRITISH TRACTOR & IMPLEMENTS PTY. LTD.
Austral Building, 95 Boundary Street, Brisbane.

86 July, 1951—POWER FARMING IN AUSTRALIA AND NEW ZEALAND

The 85mm TEA20 takes off in Australia and New Zealand. July 1951.

October, 1951 FARM MECHANIZATION 15

HELP WANTED

4 FERGUSON WORKERS FOR YOU!

Labour costs go on rising, but these Ferguson implements—all operated by one man—are worth extra men on your farm! These implements will save you labour costs from the beginning—and go on increasing your profits for years to come!

THE FERGUSON HAMMERMILL
—operated by one man!

Simple 3-point linkage to the Ferguson tractor enables the Hammermill to be taken anywhere on the farm. Look at these important factors:—

* Grinds grain, unthreshed sheaves, hay, etc. to ANY fineness.
* Fed direct from granaries, and ground material bagged on the spot or elevated 40 to 60 feet with the aid of additional tubing.
* Raised for transport by finger-tip control.

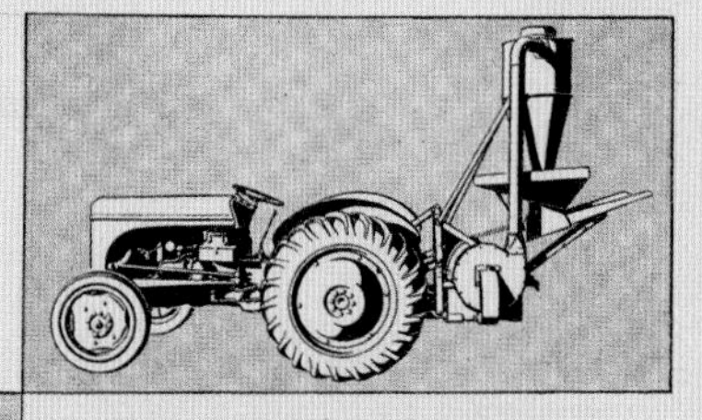

THE FERGUSON UNIVERSAL SEED DRILL —operated by one man!

The Ferguson Universal Seed Drill will constantly add to your profits! It cuts labour costs—and enables you to make full use of the right weather.

* Single Speed Drive. All adjustments of seeding rates are made by means of a lever moving over graduated dial.
* Seeds up to 7 bushels per acre for grain such as wheat and barley, down to 2 lbs. per acre for turnips and swedes.
* 13 Coulters adjustable for maximum total seeding width of 7 ft.

THE FERGUSON MANURE LOADER AND SPREADER — operated by one man!

For the first time, mechanical manure handling comes within easy reach of the small farmer!

* The loader is tractor mounted and operated hydraulically by finger-tip control.
* The spreader is easily hitched to the Ferguson tractor—without any lifting or tugging.
* One man can load, haul and spread 25 loads a day, without leaving the driver's seat. Spreading rate adjustable from 6-20 tons per acre.

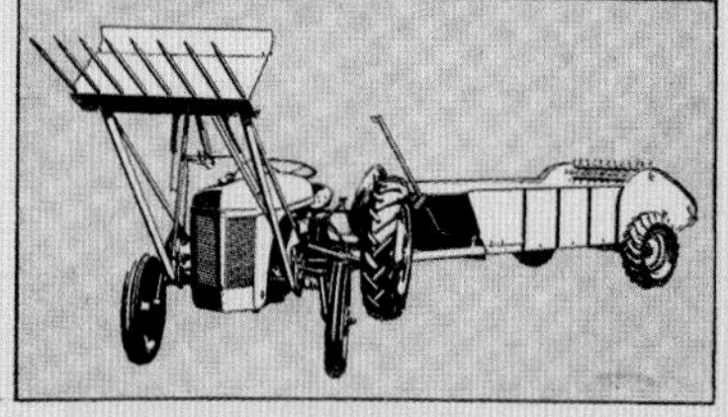

ON ANY SIZE FARM — IT WILL PAY YOU, TOO — TO

FARM WITH FERGUSON

Ferguson tractors are manufactured for Harry Ferguson Ltd., Coventry, by The Standard Motor Company Limited

One man does these jobs using the Ferguson System, where before it took more than one. October 1951.

2 FARM MECHANIZATION July, 1951

Farmers can beat shortages and rising prices

In spite of the rising cost of labour, feed, seed and equipment . . . and increasing shortage of materials, farmers can still produce more food more cheaply. But they can only do this if their production costs are lowered and their labour problems solved by the use of more economical, more versatile, more time-and-labour-saving machinery for every farming job.

The Ferguson System Saves Material, Time and Labour

The world is short of food and raw materials. The Ferguson System produces *more* of the one while using *less* of the other. In the manufacture of Ferguson tractors and implements, less iron and steel, and a greater proportion of lighter, stronger alloys is used. The Ferguson System also eliminates the need for excess built-in weight. No power is wasted. Less fuel is used. The system works on softer ground, on steeper hillsides, in more confined spaces than heavier machinery. In addition, most Ferguson implements are operated by one man from the driver's seat. They are raised, lowered, and their working depth is controlled hydraulically. Automatic protection against damage from hidden obstacles prevents breakages and costly delays. *One spanner does all necessary adjustments.*

Cheap to buy, cheap to run, and cheap to maintain, the Ferguson System offers the farmer faster, more efficient farming at a greater saving in time, material and manpower than ever before.

Grow more food-more cheaply-with Ferguson

July, 1951 FARM MECHANIZATION 3
Ferguson helps the Farmer in almost every task
Ferguson 3-Point Linkage joins tractor and implement together as one unit. Implement weight and suck of soil is transferred through the top link to the tractor's front and rear wheels. The harder the job, the greater the pressure, giving traction and stability without built-in weight.
Hitched up in 20 seconds. Unloaded in 60! With the Ferguson automatic hitch this 3-ton Hydraulic Tipping Trailer can be operated easily by one man. It will work on hillsides and on soft ground under conditions impossible to any other tractor-trailer in the world. A non-tipping model is also available.
The Ferguson Potato Spinner makes light work of potato lifting. Working as one unit with the Ferguson tractor, it has a positive drive from the power take-off. Working depth is set from the driver's seat by finger-tip and hydraulic control. A safety device protects against breakages.
There are Ferguson implements for all these jobs
Hauling · Carrying · Ploughing · Cultivating · Harrowing · Weeding · Sowing · Mowing · Loading Manure Spreading · Fencing · Planting · Shifting and Levelling Earth · Grading Roads · Digging Drains Terracing · Ridging · Potato Planting and Digging · Breaking-up Sub-Soil · Milling · Sawing · Pulling Out Trees and Roots · Trenching
Ferguson tractors are manufactured for Harry Ferguson Limited, Coventry, England, by The Standard Motor Company Limited
A3

Cheap to buy, cheap to run and cheap to maintain. Note that instead of a listing of implements there is now a listing of tasks which can be undertaken. July 1951.

2 FARM MECHANIZATION September, 1951
Out of date farming is out of pocket farming
Neither animal power nor antiquated machinery can cut man-hours and operating costs, or increase output, sufficiently to beat rising wage and production bills. This can only be achieved by the standardisation of cheaper, more versatile, more efficient machinery than ever before.
The Ferguson System joins Modern Methods to Traditional Skill
With the Ferguson System of complete farm mechanisation modern industrial methods are allied to traditional farming skill. Working as one unit with a wide variety of implements, the Ferguson tractor develops greater traction, penetration and stability without excess weight. It works on softer soil, on steeper hillsides than heavier machinery, does the same work on lower fuel consumption. It has automatic protection against damage to implements from hidden obstacles. One spanner does all the necessary field adjustments. Most implements can be operated by only one man, hydraulically from the driver's seat. Cheap to buy, cheap to run, easy to maintain, the Ferguson System enables farmers to do more work, more accurately, in less time under all conditions.
It will pay you, too–to farm with Ferguson
September, 1951 FARM MECHANIZATION 3
Ferguson helps the Farmer in almost every task
A touch of the finger can raise or lower implements hydraulically from the tractor's seat, and controls working depth.
Beat the weather with the Ferguson Universal Seed Drill! Seeds up to 7 bushels per acre of grain such as wheat and barley down to 2 lbs. per acre of turnips and swedes.
Cuts labour costs! With the Ferguson Manure Loader and Spreader, one man can load, haul and spread 25 loads a day, without leaving the driver's seat. Spreading rate adjustable from 6 - 30 tons per acre.
There are Ferguson implements for all these jobs
Hauling · Carrying · Ploughing · Cultivating · Harrowing · Weeding · Sowing · Mowing · Loading Manure Spreading · Fencing · Planting · Shifting and Levelling Earth · Grading Roads · Digging Drains Terracing · Ridging · Potato Planting and Digging · Breaking up Sub-Soil · Hoeing · Milling · Sawing Pulling Out Trees and Roots · Trenching
Ferguson tractors are manufactured for Harry Ferguson Limited, Coventry, England, by The Standard Motor Company Limited

The standardisation of cheaper and more efficient machinery than ever before to cut costs and increase output. September 1951.

By 1952 Ferguson is seen to have developed 80 different implements to work with his Ferguson tractor. June 1952.

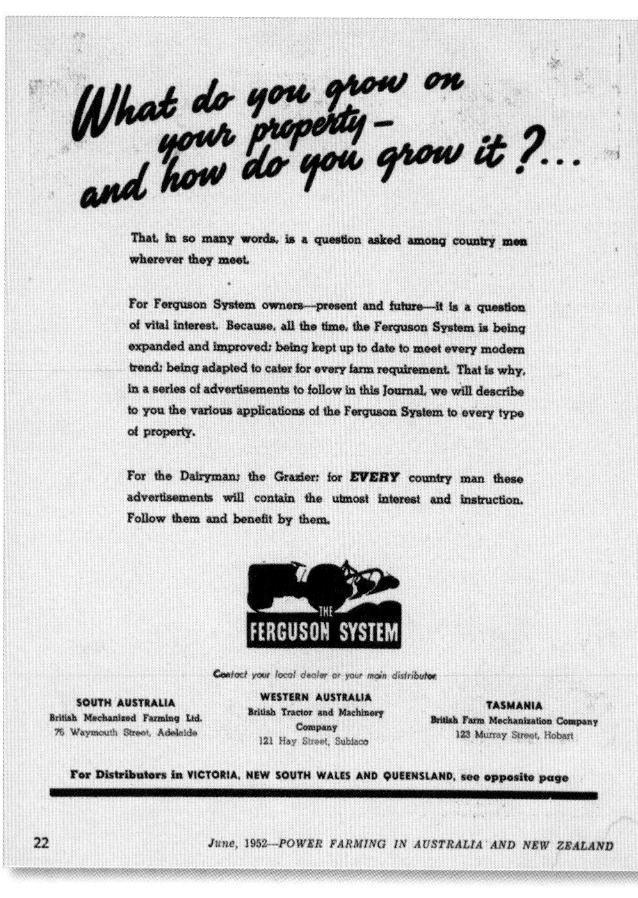

What do you grow on your property – and how do you grow it ?...
That, in so many words, is a question asked among country men wherever they meet.
For Ferguson System owners—present and future—it is a question of vital interest. Because, all the time, the Ferguson System is being expanded and improved; being kept up to date to meet every modern trend; being adapted to cater for every farm requirement. That is why, in a series of advertisements to follow in this Journal, we will describe to you the various applications of the Ferguson System to every type of property.
For the Dairyman; the Grazier; for EVERY country man these advertisements will contain the utmost interest and instruction. Follow them and benefit by them.
THE FERGUSON SYSTEM
Contact your local dealer or your main distributor
SOUTH AUSTRALIA British Mechanized Farming Ltd. 76 Waymouth Street, Adelaide
WESTERN AUSTRALIA British Tractor and Machinery Company 121 Hay Street, Subiaco
TASMANIA British Farm Mechanization Company 123 Murray Street, Hobart
For Distributors in VICTORIA, NEW SOUTH WALES AND QUEENSLAND, see opposite page
22
June, 1952—POWER FARMING IN AUSTRALIA AND NEW ZEALAND

FOR EVERY SEASON . . .
SPRING
AUTUMN
WINTER
SUMMER
FOR EVERY PROPERTY
. . . there's a Ferguson System Implement for your EVERY farming job!
An implement and a tractor operated as one unit by ONE MAN.
Think what this means to you. In the Ferguson System there is a tractor—a Ferguson Tractor—with over 80 different implements to work with it.
Naturally, on a single property, you won't find a use for 80 implements. But, there will be several which will enable you to work your property—better, faster. Implements which, when attached to your Ferguson tractor, are operated by you—or by any ONE MAN—from the seat of the tractor.
Here are some of the jobs which can be done with one Man and a Ferguson: • Ploughing • Cultivating • Harrowing • Mowing • Manure spreading • Trenching and draining • Potato planting and digging • Sub-soiling • Fencing • Terracing • Weeding • Sowing • Earth shifting and levelling • Grading • Loading, hauling, carrying • Milling • Sawing • Tree and root removal.
For fuller information on the Ferguson System, contact your local dealer or main distributor.
VICTORIA British Farm Equipment Co. 568 Elizabeth Street Melbourne
NEW SOUTH WALES British Farm Equipment Co. 83-97 Flinders Street Sydney
QUEENSLAND & N.T. British Tractor and Implements Pty. Ltd. Boundary Street, Brisbane
FERGUSON SYSTEM
For Distributors in S.A., W.A. and TAS., see opposite page
POWER FARMING IN AUSTRALIA AND NEW ZEALAND—June, 1952
23

Ferguson goes poetic in colour again for Christmas. December 1951.

Two Shillings and Sixpence
December, 1951
British Farm Mechanization
THE LINK BETWEEN FARMING & ENGINEERING
HARRY FERGUSON LTD., COVENTRY
Ferguson Tractors are manufactured for Harry Ferguson Ltd., Coventry, by The Standard Motor Co., Ltd.

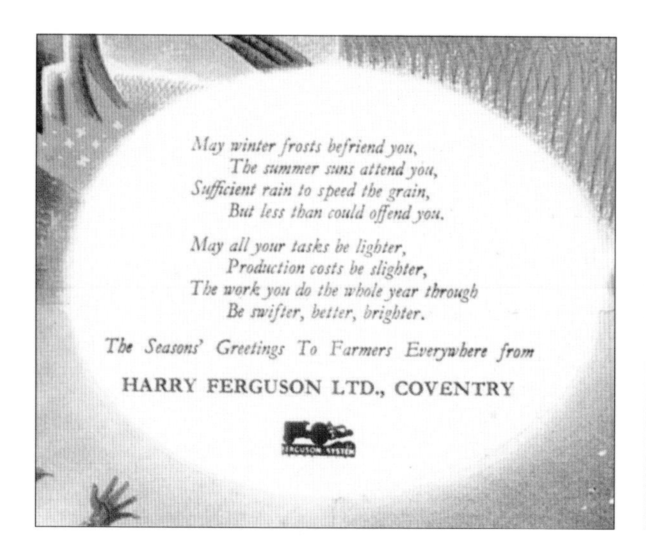

May winter frosts befriend you,
The summer suns attend you,
Sufficient rain to speed the grain,
But less than could offend you.
May all your tasks be lighter,
Production costs be slighter,
The work you do the whole year through
Be swifter, better, brighter.
The Seasons' Greetings To Farmers Everywhere from
HARRY FERGUSON LTD., COVENTRY

During the last five years over one hundred and fifty thousand famers have bought two hundred and thirty thousand British-made Ferguson tractors and many times that number of British-made Ferguson implements. And our own satisfied customers have been the best of all Ferguson salesmen.

When one farmer buys a Ferguson his neighbours follow suit. They are able to see for themselves how the Ferguson System saves time, labour and money under all conditions. They are investing in a proved product.

The past five years have shown a steady advance—not only in Ferguson sales, but in the development of a fundamentally brilliant conception and design. These developments have produced easier starting, greater fuel economy, less engine wear, and increased power. They have also produced a wider range of tractors and implements to meet more and more farming needs.

But throughout these five years, the fundamental design of the Ferguson System has remained unchanged, because it withstands the test of time. It has proved beyond question that the Ferguson tractor-implement System produces more traction with less weight, uses less material for construction and less fuel in operation, enables fewer men to do more work, more cheaply, more accurately, and more easily than ever before.

On any size farm it will pay you too to farm with FERGUSON

Ferguson tractors are manufactured for Harry Ferguson Ltd., Coventry, by The Standard Motor Company Ltd.

Very strange that one year on from the first claim of 150,000 tractors made in December 1950, the figure is still the same! Serial numbers indicate that over 240,000 had been made by this time. Did someone get it wrong? January 1952.

March, 1952 FARM MECHANIZATION 15

More Farmers choose Ferguson than any other tractor

In 5 years the Ferguson System has become Britain's most popular farm equipment. Now, despite a production rate of 300 tractors a day, there are still more needed than can be made. That is what happens when machinery truly fulfils its function to increase production and lower production costs.

For this remarkable success it is the Ferguson System —the union of Ferguson tractors with Ferguson implements—that is primarily responsible.

For it is the System—with its unique 3-point linkage and hydraulics—that eliminates the need for wasteful, engine-straining, soil-packing weight. It is the System that allows for simple, quick and safe interchange of implements. It is the System that protects implements from breakage. It is the System eliminating needless weight, that brings you tractor and implements at a low price. And it is the System that, above all, will grow you more food at less cost—which is what every farmer seeks and the country needs.

When you buy the Ferguson System you are investing in a proven product. No wonder farmers prefer it !

ON ANY SIZE FARM IT WILL PAY YOU, TOO—TO Farm with Ferguson

FERGUSON SYSTEM

Ferguson tractors are manufactured for Harry Ferguson Ltd., Coventry by The Standard Motor Company Ltd.

The Ferguson at the height of its popularity. March 1952.

April, 1952 FARM MECHANIZATION 5

CUT WORKING COSTS AS WELL AS GRASS

WITH THE FERGUSON TRACTOR-MOUNTED MOWER

You'll have easier, swifter mowing, with fewer delays than ever before with the Ferguson Mower. Tractor-mounted, power operated, it cuts a 5 ft. swathe at speeds up to 3½ m.p.h. in second gear under all conditions.

The knife speed of the Ferguson mower is considerably above average, and thus deals more efficiently with the short type of grass cut for silage and grass drying. A break-away mechanism in the pull bar protects the cutter bar from damage on hitting obstacles, and avoids costly delays on uneven ground. The cutter bar is located just behind the tractor rear wheels, and this, combined with the use of the hydraulic lift and independent rear wheel brakes gives the quickest turns and highest degree of manoeuvrability you've ever found on a mower.

The mower can be attached and detached from the tractor by one man in a matter of minutes. It's one of the finest Ferguson aids to more efficient, more economical farming.

Ask your local Ferguson Dealer for a free demonstration on your farm and about the Ferguson Pay-as-you-Farm Plan.

ON ANY SIZE FARM — IT WILL PAY YOU, TOO — TO

FARM WITH FERGUSON

Ferguson tractors are manufactured for Harry Ferguson Ltd., Coventry, by The Standard Motor Company Limited

A contented farmer, composed and relaxed with that mower which was so tedious to hitch! April 1952.

May, 1952 FARM MECHANIZATION 9

"KEEP THAT MOWER GOING!"

How the Ferguson System set a new style in grass management

90 almost derelict acres transformed into first-class pasture! When Capt. Coles started, ten years ago, to improve the condition of this Cotswold estate, park and woodland were a mass of thistle, hawthorn and scrub. Now, through careful grazing and intelligent use of the versatile Ferguson System, the land has been turned into healthy sward of such high feeding value that stock are fattened on it without any concentrates whatsoever.

Capt. G. Coles, of Campden Farm Estates Ltd., outside his house near Chipping Campden, Glos.

With the Ferguson tractor-mounted Power Mower, one man can cut a 5 ft. swathe at 3½ m.p.h. Automatic protection against hidden obstacles. Easily detached.

Capt. Coles "cannot speak too highly" of the Ferguson System and how it has helped him to grow more food so cheaply:—

"The Ferguson tractor and mower . . . will go where no other mower can possibly work without turning over. It will cut banks and steep places at an angle of 45 degrees. It will improve the old pastures beyond recognition. Simply by cutting all the coarse grasses it enables the young succulent grasses to grow. It helps to produce a sweet and palatable feeding for sheep and cattle the whole time.

"It will stand up to terrifically hard work. On these Cotswold hills, as everyone knows, stones are very often found above ground and the going is very rough, but still the mower stands up to it.

"I have had this Ferguson mower since it first came out and I cannot speak too highly of it—therefore, my advice is 'keep that mower going' and keep your pastures clean."

(The original letter can be inspected at the offices of Harry Ferguson Ltd., Coventry.)

ON ANY SIZE FARM

Ask your nearest Ferguson Dealer for a demonstration on your farm—now. Ask him, too, about the Ferguson Pay-as-you-Farm Plan.

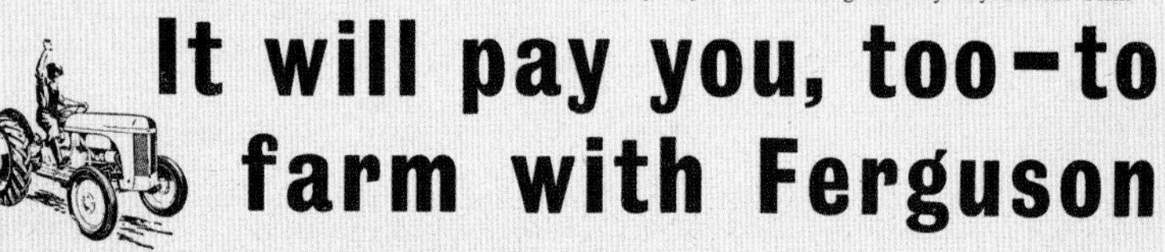

It will pay you, too–to farm with Ferguson

Ferguson tractors are manufactured for Harry Ferguson Ltd., Coventry, by the Standard Motor Company Ltd.

81

Improve your steep grassland with a Ferguson mower – the Captain can't be wrong. But in a more serious vein, one of the Ferguson tractor's greatest attributes was its stability on slopes – possibly unrivalled by any other two-wheel drive tractor. May 1952.

The mighty power of a great new engine
plus the one and only Ferguson System
...THAT'S THE
FAR MORE POWERFUL
FERGUSON Twenty 85
Ferguson's mighty valve-in-head engine, with the most remarkable torque characteristic of all, gives you enormous lugging ability. This power, working with the genuine Ferguson System, gives unexcelled performance that only the Ferguson Twenty-85 can offer.
The important thing to be said here is that performance such as this permits you to farm more of your land . . . easier, more intensively and at lower cost. You'll find, as have so many other owners, that the Ferguson Twenty-85 meets more of your needs more of the time than any other tractor.
There's much more to the story of Ferguson Power and the Ferguson System, of course, and any Ferguson Dealer will gladly give you all of the details.
GET "SHOWDOWN" PROOF
Call your Ferguson Dealer today and arrange for a "Showdown" demonstration right on your own farm. Then compare Ferguson all-around performance with that of any other tractor.
★
Ask your Ferguson Dealer for a copy of "The Axe and the Wrench" . . . a simple explanation of torque. Or write to Harry Ferguson, Inc., Detroit 32, Michigan.
Copyright 1952 by Harry Ferguson, Inc.
Meets MORE of the needs of MORE of the farmers MORE of the time!
MAY 1952
5

UK built Ferguson tractors exported to Canada were called the Ferguson Twenty-85. They had the 85mm version of the original 80mm Standard engine. May 1952.

June, 1952
FARM MECHANIZATION
17
Grow more food, more cheaply with Ferguson
Now, as never before, the nation needs more food at less cost. The farmer needs to achieve still bigger savings of time, labour and money in order to reach this goal.
The Ferguson System—the simplest form of mechanised farming ever known—has proved to be the successful way of getting more from the land more cheaply. This versatile System consists of a light, powerful tractor working as one unit with a wide range of implements. It produces greater traction and stability without massive built-in weight. It is cheap to buy, cheap to run—fuel costs are exceptionally low. It is so easy to use that anyone—old or young—can operate it.
The Ferguson System is today helping farmers in 76 countries to farm better . . . faster . . . and cheaper. Ask your Ferguson Dealer for a demonstration on your farm; ask him, too, about the Ferguson Pay-as-you-Farm Plan.
There is a Ferguson implement for each of these jobs!
Irrigating · Hauling · Ploughing · Cultivating · Sowing
Mowing · Transporting · Harrowing · Weeding
Loading · Manure Spreading · Fencing · Planting
Shifting and Levelling Earth · Grading Roads
Terracing · Ridging · Potato Planting and Lifting
Sub-Soiling · Hoeing · Milling · Sawing · Pulling Out
Trees and Roots · Belt-work
ON ANY SIZE FARM—
IT WILL PAY YOU, TOO—TO
Farm with Ferguson
Ferguson tractors are manufactured for Harry Ferguson Ltd., Coventry, by The Standard Motor Company Ltd.
cl

Was this a vision of things to come? Is it a Massey-Harris binder? Notice the registration number of the tractor is HF 1952 – fact or fiction?

July, 1952 FARM MECHANIZATION 9

Bigger savings all round the farm with Ferguson

Now, when farmers must cut costs to a minimum, the advantages of the versatile Ferguson System are being more and more widely appreciated. Exceptionally low running costs, low all-inclusive purchase price, ease of handling and ability to speed every job—these are only a few of the factors which make more and more farmers choose Ferguson!

Ferguson 3-Ton Hydraulic Tipping Trailer, shown above, can be quickly coupled and uncoupled—without leaving the tractor seat—by automatic hydraulic hitch. Fingertip control, too, for tipping—to 40 degrees. Easy to handle, cheap to run, goes anywhere—up steep hills, over sticky ground.

Ferguson Hammermill produces stock feed, etc., quickly, easily—anywhere you want. Transported by hydraulic control on tractor or can be easily adapted as a fixture in barns, etc. Belt-drive gives high output at low cost. Easily changeable milling screens. Ground material bagged on the spot.

Ask your Ferguson Dealer for a demonstration on your farm; ask him, too, about the Ferguson Pay-as-you-Farm Plan.

HARRY FERGUSON LTD.

Ferguson tractors are manufactured for Harry Ferguson Ltd., Coventry, by The Standard Motor Company, Ltd.

B1

Linking the functionality of two Ferguson implements? The Ferguson trailer carrying grain to the Ferguson mill. July 1952.

September, 1952 FARM MECHANIZATION 17

"Ferguson servicing, too, is revolutionizing farming!"

—says Yorkshire farmer, Mr. R. Pumphrey

Bishop Keld Farm,
Pateley Bridge,
Harrogate.

Messrs. Glovers of Ripon Ltd.,
Ferguson Service Department,
RIPON.

Dear Sirs,

Your "Field Service" for Ferguson tractors is well-known, but I must say how extremely gratified I am by the swift manner you are able now on the farm to undertake repairs that are more extensive than any day-to-day adjustments.

I had been given reasonable expectation that my tractor would require a day's attention on this job . . . and to find that this could be done in my own workshop and completed within the forenoon, through your normal servicing unit, leaves one extremely gratified and well impressed.

We were one of the first in Nidderdale to have a Ferguson and it has never been more in use than during the past year, during which, in addition to usual duties, it has enabled us to ensile 35 tons of Arable Silage . . . without need of our calling in outside labour.

To my mind, there can be no doubt that not only the Ferguson but Ferguson Servicing is revolutionizing farming in this Dale country to a far greater extent than is generally realised.

Yours very truly,
(sgd.) Roland Pumphrey

Further proof of the success of the versatile Ferguson System comes in this unsolicited letter.

Ferguson Service starts from the very moment a Ferguson tractor or implement is delivered to the farm. Every Ferguson tractor is given six months' free service by factory-trained engineers and regular follow-up calls are made to ensure maximum efficiency in operation of both tractor and implements. Special time-saving tools, immediate and skilled attention are among the reasons why the Ferguson owner can depend on the Ferguson Organisation to help him produce more food at less cost.

Now see what this satisfied Ferguson farmer has to say!

Ask your nearest Ferguson Dealer for a demonstration of the unique Ferguson System on your farm—now. Ask him, too, about the Ferguson Pay-as-you-Farm Plan.

Ferguson

PIONEERS OF ON-THE-FARM SERVICE

Ferguson tractors are manufactured for Harry Ferguson Ltd., Coventry, by The Standard Motor Company Ltd.

C1

Ferguson Servicing also revolutionises farming. September 1952.

FARMER and STOCK-BREEDER, November 11-12, 1952
How Ferguson slashes farming costs
By lower purchase price...
For the low inclusive price of £395*, you can buy a Ferguson tractor equipped with these unique outstanding features: built-in hydraulic mechanism, 3-point linkage, p.t.o., safety electric starter, pneumatics, automatic protection against obstructions, adjustable wheel widths, adjustable drawbar.
By lower running costs...
The design of the Ferguson System gives traction without built-in weight, makes a light tractor do a heavy tractor's work at low running cost. Unequalled range of Ferguson implements — "tailored" for the tractor and for the work required — give increased yields, again at less cost.
By lower servicing costs...
The design and high-grade materials used in the manufacture of Ferguson tractors and implements, together with Ferguson "on-the-farm" service, guarantee long and trouble-free life.
That's how Ferguson slashes farming costs!
The Ferguson System
Ask your Ferguson Dealer for a free demonstration on your farm; ask him, too, about the Ferguson Pay-as-you-Farm Plan.
(*Ex works)
FERGUSON TRACTORS ARE MANUFACTURED FOR HARRY FERGUSON LTD., COVENTRY BY THE STANDARD MOTOR COMPANY LTD.

The price might now be £395 for a petrol tractor, but the Ferguson Pay-as-you-Farm Plan is by now well established and there for the asking. November 1952.

December 19, 1952 THE RAILWAY GAZETTE 25

NOW—The Ferguson Industrial Tractor

Petrol, vaporising oil or diesel models.
One all-inclusive price covers: built-in fingertip hydraulic control, power take-off, two independent brake systems, hinged mudguards, safety starter, spring-loaded bumper, tipping seat, lights, horn and driving mirror.

Here's a new, economical, mechanical handyman! This new Ferguson is the versatile, industrial brother of the hundreds of thousands of agricultural Fergusons now slashing costs on the land! It's easy to handle and maintain, cheap to buy and run. It's powerful, compact and manœuvrable. It can go almost anywhere, and with the unique range of Ferguson implements, do almost anything!

See for yourself how it can save time, money and labour in *your* works. Consult your nearest Ferguson Dealer; or write direct for further details to Harry Ferguson Ltd., Industrial Tractor Division, Coventry.

THESE are only a few of the many jobs this new Ferguson does better, faster, cheaper!

- All types of haulage.
- Winching.
- Post hole digging.
- Grass mowing.
- Shunting trucks and wagons.
- Shifting and levelling earth.
- Loading and unloading.
- Stationary belt-work.
- Grading roads.

HARRY FERGUSON LTD.

Ferguson tractors are manufactured for Harry Ferguson Ltd., Coventry, by The Standard Motor Company Ltd.

Having dealt with the farmers of the world, let's now deal with industry! A rare advertisement for an Industrial Ferguson found in a Railway magazine. December 1952.

Years after the development of the basic Ferguson System concept, the "System" message still had to be repeated. The price of the petrol tractor has by now increased to £395. November 1952.

A comprehensive mechanisation package for farmers down under. December 1953.

No poetry this Christmas. Just get on with muck shifting on your own, thanks to the Ferguson System. Even the Wellington boots are not needed because you can load, hitch and spread without leaving the seat. December 1952.

Break your heavy dry lands with a Ferguson disc plough. April 1953.

Ferguson helps the dairy farmer by doing all the jobs between milking times. January 1953.

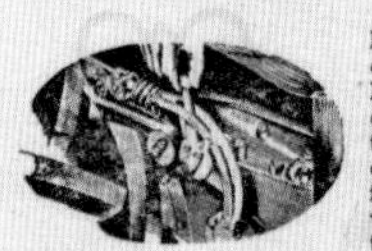

The Ferguson "System" was exclusive to Ferguson tractors. The advantages of the System reiterated. June 1953.

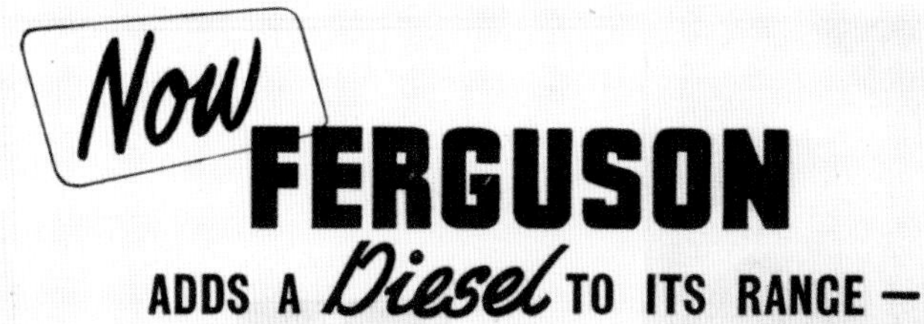

ADDS A *Diesel* TO ITS RANGE –

..to give you
your choice
of Three
fine models
PETROL
KERO
DIESEL

★ Four cylinder O.H.V. Diesel engine developing 27 B.H.P.

★ Instantaneous starting on Diesel fuel

★ Operates on Automotive Diesel fuel (distillate)

★ Special patented combustion chamber for quiet, smooth operation

★ Replaceable Cylinder Liners

★ Pre-engaged type starter motor

★ Thermostatically controlled cooling system

★ AFTER SALES SERVICE

An Australia-wide Dealer Organization provides complete after-sales service and spare parts facilities

FERGUSON—The Diesel With a Difference

All the proven advantages of the Ferguson System plus the extra economy of Diesel Operation

IT'S HERE — NOW ! A tractor designed by Harry Ferguson Ltd. to bring all the advantages of the Ferguson System to farmers who require Diesel power

In addition to the established economy of diesel operation, this useful, versatile model brings all these Ferguson advantages to the diesel field . . . Automatic depth control, Built-in hydraulic system, Penetration without weight, Three-point linkage, Traction without in-built weight, Protection from hidden obstructions and Adjustable wheel tracks

Truly, it's the diesel with a difference. See your Ferguson dealer now, or mail coupon at left, for prices, details, recommended uses, &c. Enquiry entails no obligation

BRITISH FARM EQUIPMENT Pty. Ltd.

568-576 Elizabeth St., Melbourne, C.I

Dear Sir: Please send me full particulars of your Ferguson Tractors (Mark X alongside the type, or types, in which you are interested)

☐ DIESEL ☐ PETROL ☐ KEROSENE

NAME ....

ADDRESS ....

.... VJA 10

See Your Dealer or Mail this Coupon

BRITISH FARM EQUIPMENT

PTY. LTD.

568-576 ELIZABETH STREET, MELBOURNE, C.I

Telephone : FJ 2154

Dealers and Service everywhere throughout Australia

Ferguson Tractors are manufactured by the Standard Motor Co. Ltd. for Harry Ferguson Ltd., Coventry, England

The Journal of Agriculture, Victoria—October, 1953

Harry Ferguson was a late convert to diesel engines but its introduction kept the tractor as popular as ever and reduced operating costs for farmers as cheaper diesel fuel became increasingly available. October 1953.

FARMER and STOCK-BREEDER, May 19, 1953

At the service of the Queen's yeomen . . .

Stoneleigh Abbey, Warwickshire. Built by Sir Thomas Leigh, Lord Mayor of London in the first year of the reign of the first Elizabeth, by whom he was knighted . . . Today, this stately mansion continues to play a vital part in the nation's progress under the second gracious Queen of that name . . .

By kind permission of the present Lord Leigh, part of the estate is now occupied by the Ferguson Mechanised Farming School. Through the rapidly developing educational system centred on this School, the Queen's yeomen —the landowners, farmers and smallholders of Britain —are acquainted with the simplest, most effective and most economical method of farming ever known Through this famous Ferguson System, their labour and land is rendered more productive, their future more secure.

Through its fruitful effects on land of every sort, on farms of every size, the Ferguson System is helping progressive landowners and farmers to contribute in ever-growing degree to the nation's agricultural production . . . keeping in good heart the lands over which our new Queen rules . . .

Loyal Greetings to Her Majesty!

FERGUSON TRACTORS ARE MANUFACTURED FOR HARRY FERGUSON LTD., COVENTRY, BY THE STANDARD MOTOR COMPANY, LTD.

Harry Ferguson had his first Training School in the UK in a part of the Stoneleigh Abbey complex of buildings. Then it moved to "modern" ex US air force hospital buildings that had been built on the Stoneleigh Estate during World War II. It was known as the Ferguson Mechanised Farming School later becoming part of the Massey Ferguson empire. Thousands of UK and overseas students went through this school. May 1953.

Much advertising in early 1953 concentrated on how the Ferguson System could cut costs. Eight good line drawings emphasise the range of tasks that the Ferguson can undertake. January 1953.

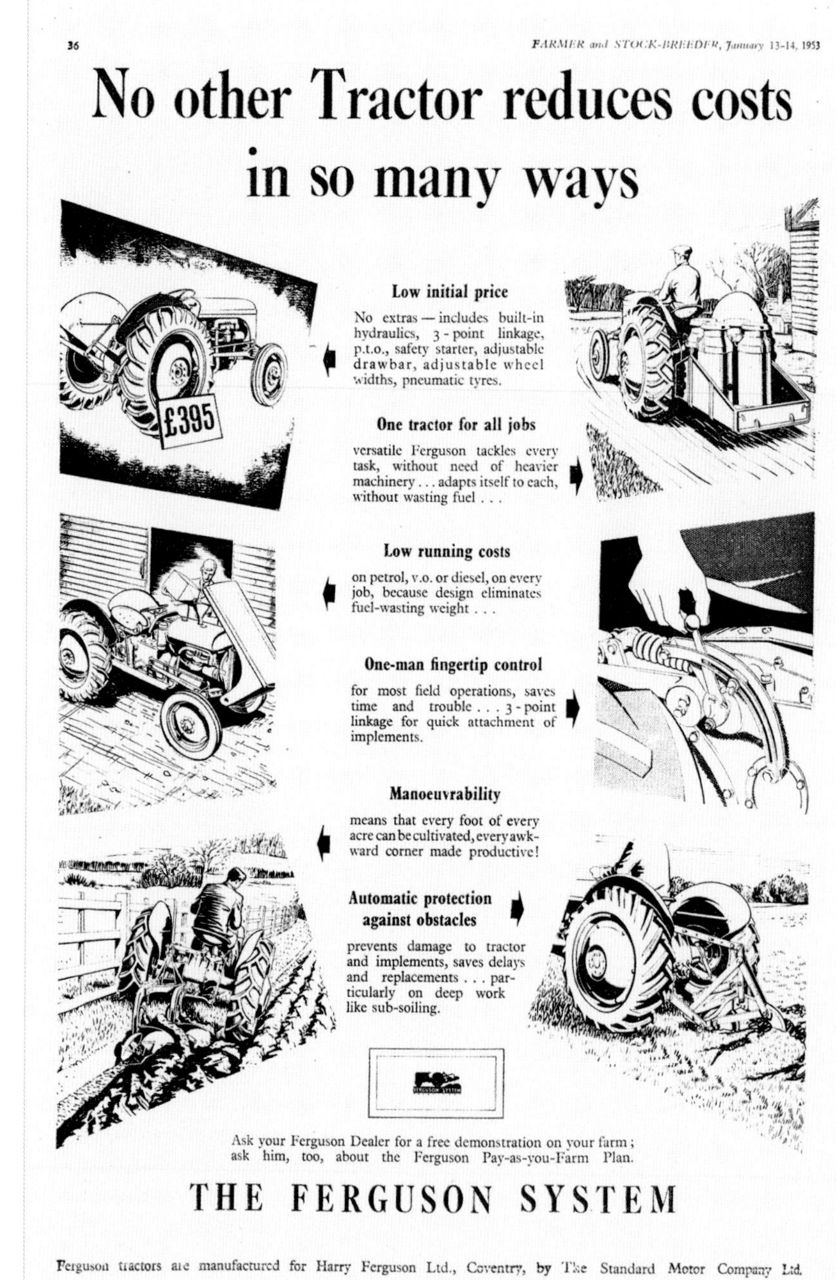

This appears to be one of the first advertisements containing reference to the availability of a diesel engine version of the Ferguson tractor for the first time in either UK or North America. 1953

More good line drawings of implements to bring home the versatility and economy of the Ferguson System. April 1953.

56 The Farmers Weekly, October 23, 1953

The Ferguson ploughing three furrows up a 1 in 3½ hill.

Ploughing up steep hills

It is a fallacy that bigger, heavier wheeled tractors can beat the Ferguson ploughing up steep hills.

Heavy wheeled tractors need most of their power to propel their great weight uphill, leaving little to spare for work. Their implements add still more wasteful weight.

Not so with the Ferguson.

Because of its unique design, the Ferguson eliminates built-in weight, transfers the natural forces of the implement in work to add penetration and traction. This gives perfect performance under all conditions, saves fuel and provides power to spare for the steepest hills.

Prove it for yourself by asking the salesman who claims his tractor is the equal of the Ferguson, to meet the Ferguson in open competition anywhere on the steepest hills. Also, insist that a full competitive demonstration be made with a full range of equipment. This will prove beyond any doubt our claim that the Ferguson meets more of the needs of all of the farmers, for more jobs, more economically, than any other tractor.

Harry Ferguson Ltd.

Ferguson tractors are manufactured for Harry Ferguson Ltd., Coventry, by The Standard Motor Company Ltd.

In 1948 we saw the Ferguson pulling a two furrow plough up a 1 in 2.5 hill. Now things have moved on to three furrows being pulled up a 1 in 3.5 hill. Maybe this was a sign of the times – Ferguson had by now merged (on 26th August 1953) with Massey-Harris to form Massey-Harris-Ferguson. The Harry Ferguson name lingers on the advertisements for a while, pending the new company getting all the stationery changed! October 1953.

50 The Farmers Weekly, October 9, 1953

Faster, more accurate sowing and fertilizing!

THE FERGUSON MULTI-PURPOSE SEED DRILL WITH FERTILIZER ATTACHMENT

One-man Ferguson farming cuts costs and time! All Ferguson implements, through unique design and materials, are tough yet light — fit for lifetime work. Here, for the urgent demands of seed - time, is the Ferguson Seed Drill with fertilizer attachment. Test this outstanding implement for yourself and see how it speeds the work, accurately, economically. Note all these features, too . . .

SEED DRILL

1. Sows most seeds in common use. No need for different drill for roots.
2. No operator required on drill. Feed visible from tractor seat. Coulters raised, lowered, by fingertip control.
3. External force feed for fast, accurate sowing over a wide variety of rates per acre—from 7 bushels cereals to 2 lbs. turnips or sugar beet. High speed gear available giving double sowing rates.
4. Overall width under 9 ft. for easy passage through gates, etc., yet covers 13 rows when sowing.

FERTILIZER ATTACHMENT

1. Easily and quickly attached to drill.
2. Completely separate unit with own feeding mechanism, which can be dismantled for cleaning in a few minutes.
3. Feed mechanism design ensures uniform flow of fertilizer at application rates of 110—1,200 lbs. per acre.
4. Corrosion-resisting low loading hopper tips to horizontal for rapid emptying and cleaning.
5. All gears and bearings self-lubricating.

Ask your Ferguson Dealer for a free demonstration on your farm. Ask him, too, about the Ferguson Pay-as-you-Farm Plan.

Harry Ferguson Ltd.

Ferguson tractors are manufactured for Harry Ferguson Ltd., Coventry, by The Standard Motor Company Ltd.

The Ferguson drill had a good reputation for reliability and accuracy. Unlike most other drills of the day there was no provision for an attendant to ride on the drill. October 1953.

Here's how to buy a tractor

Just ask yourself: "Will this tractor do the most jobs for *me, more* of the time, without wasting power? Is it easy to operate, and will it go from one kind of a job to another, quickly and easily?"

To *answer* yourself, you've got to *see* the tractor do *your* jobs on your farm. And above all, you shouldn't let "habit" guide your choice.

This time, call your *Ferguson* Dealer. Ask him to prove the ability of the Ferguson "*Twenty*-85" in a Showdown Demonstration on your farm. Let him show you how many bottoms it will pull in *your* soil. See how quickly and easily you, or anyone, can change implements with Ferguson's *time-proved* 3-point hookup. Then disc, or do any of the other jobs you'll be doing throughout the year.

You (and your family) will discover that the exclusive Ferguson System gives you a lot more tractor for a lot less money . . . a lot more farming, with a lot less work.

Call your Ferguson Dealer today. Set up a Showdown Demonstration on your farm *soon.* Chances are, you've already missed too much . . . too long!

FREE BOOKLET tells you "How to Buy a Tractor". 24 pages of valuable information! Your Ferguson Dealer has your copy, or write: Harry Ferguson, Inc., Detroit 32, Michigan.

Get your Showdown Demonstration of the Ferguson Manure Spreader and Loader. This exclusive Ferguson combination lets you load, hitch, haul and spread without leaving the tractor seat! Hydraulically operated, patented hook 'n' eye hitch lets you do this tough job alone, without drudgery.

No other tractor gives you *all* the Ferguson System advantages: Traction and penetration without power-stealing weight, finger tip and automatic draft control, front-end stability, and an exclusive built-in hydraulic overload release that saves tractor and implement if you hit a hidden rock or stump.

Seeing is Believing — Get Your Showdown Demonstration of the

FERGUSON *Twenty*-85

The UK-produced Ferguson Twenty-85 on the Canadian market. Note the five section harrows that were apparently only available in North America and which needed a stabilising tie back to the front axle. 1953.

"Industrialise" the Ferguson and see what a range of off-farm jobs it can perform. December 1953.

16 THE RAILWAY GAZETTE December 18, 1953

Another good job done at less cost than before!

Versatility pays—and here's the new Ferguson Industrial Tractor to prove it . . . doing a job normally needing much heavier, more expensive machinery! One man and one tractor can do dozens of different tasks a day—and, in doing them, slash your costs, for this Ferguson with its range of implements, works quickly, easily, economically! See, for example, how it speeds all these jobs . . .

- ★ All types of haulage.
- ★ Post hole digging.
- ★ Shunting trucks and wagons.
- ★ Loading and unloading.
- ★ Winching.
- ★ Grass mowing.
- ★ Grading roads.
- ★ Shifting and levelling earth.
- ★ Stationary belt work.

Low Purchase Price

Remember this, too—the Ferguson is real value. One all-inclusive price covers: built-in fingertip hydraulic control, power take-off, two independent brake systems, hinged mudguards, safety starter, spring-loaded bumper, tipping seat, lights, horn, and driving mirror.

Low Running Costs

Powerful, yet economical, the Ferguson Industrial Tractor is exceptionally light on fuel (petrol, vaporising oil, or diesel models). On every job, from haulage to beltwork, the smooth running Ferguson engine keeps costs down to a minimum.

Consult your nearest Ferguson Dealer; or write direct for further details to Harry Ferguson Ltd., Industrial Tractor Division, Coventry.

FERGUSON INDUSTRIAL TRACTOR

Ferguson tractors are manufactured for Harry Ferguson Ltd., by The Standard Motor Company Ltd.

32 THE RAILWAY GAZETTE February 19, 1954

Cut turn-round time!

The Ferguson Industrial Tractor. Petrol vaporising oil or diesel engine. Two independent brake systems (one hydraulic), hinged mudguards (easily detachable), power take-off, built-in fingertip hydraulic control, safety-starter, spring-loaded bumper, tipping seat. Fitted with lights, horn and driving mirror for road work.

One man and one Ferguson tractor can save time and money on dozens of different jobs a day. Hauling loads in the yards . . . loading and unloading at the sidings . . . shunting trucks and wagons . . . scarifying railway tracks . . . shifting earth . . . belt-work—these are only a few of the versatile new Ferguson's many uses. It handles easily in confined spaces. It's cheap to buy and run. It can go almost anywhere, and, with the unique range of Ferguson implements, do almost anything—better, faster, cheaper.

See for yourself how this tractor saves time, money and labour. Massey-Harris-Ferguson will gladly arrange a free demonstration for you any time.

Ferguson Industrial Tractor

Ferguson tractors are manufactured for Massey-Harris-Ferguson, Coventry, by The Standard Motor Company Ltd.

How Ferguson tractors could give British Railways a helping hand. February 1954

ADC – Automatic Draft Control – now becomes the buzz phrase for salesmen. And all explained in a handy free Ferguson booklet – a collectors item now. May 1954.

Many more Ferguson earth scoops were sold down under than in the UK. The earth scoop advertised with other non-arable implements. August 1954.

Sounds absurd but go fishing with your Ferguson! And bring the catch home on that now exceptionally rare model of trailer – the Ferguson 2 ton which was only made down under. November 1954.

1954 brought in the Massey-Harris-Ferguson company name and Ferguson effectively becomes a brand name alongside Massey-Harris equipment as the new company pursued a two line policy. April 1954.

This splendid colour advertisement is based on an original painting by Terence Cuneo who was commissioned to paint several such scenes for Massey-Harris and clearly he ran over into the M-H-F period. The picture clearly shows what the merger of the two companies was all about – Ferguson's strength in tractor design and Massey-Harris's strength in harvesting. July 1954.

July, 1954 FARM MECHANIZATION 27

The sums are now getting better and more accurate! This figure reflects total production of Ferguson System type tractors in both UK and USA. July 1954.

Why farmers have bought 820,000 Ferguson tractors!

"A farming revolution!" That's what farmers said when the Ferguson System arrived. It is still as true today. Proof of the System's remarkable success is in the fact that over 820,000 Ferguson tractors and over 2,500,000 implements are now in use. Farmers everywhere are now farming with Ferguson—and finding that it pays handsomely!

They have found, too, that only the Ferguson System incorporates all the vital features others have tried to imitate . . . features found in their entirety only in a Ferguson. Only Ferguson produces traction without fuel-wasting weight, front-end stability and built-in protection against damage from obstruction.

Only Ferguson has such simple maintenance, quick implement changes, one-man operation of so many jobs and extra-special manoeuvrability!

These are not mere claims. These are the facts of proved performance . . . performance tried and tested in the field, in all soil conditions, on farms of every size. Ask the man who knows—the Ferguson farmer!

FREE DEMONSTRATION! Your Ferguson Dealer will prove these facts to you in a free demonstration on your own land, on any day you name . . . without obligation! He'll tell you, too, about the Ferguson Pay-as-you-Farm Plan, which makes immediate Ferguson ownership so easy! See him today!

Ferguson

See us at the 'Royal'—Stand No. 241

Ferguson tractors, sold by Massey-Harris-Ferguson (Sales) Ltd., are manufactured by The Standard Motor Co. Ltd., Coventry.
D3

Fast haymaking and transport for Australian farmers. November 1954

December, 1954 FARM MECHANIZATION 41

SEE HOW THIS TRACTOR WORKS AND WORKS AND WORKS!

No limit to the versatility of the Ferguson tractor. It gets more work done, cuts more costs on any size farm — your farm! How? Mechanise your farm-work with the Ferguson System and see! This System of hydraulics and three-point linkage transfers the weight and force of the implement in work on to the tractor. This gives you extra power, penetration and traction whenever and wherever you need it. In fact, you get two tractors in one! A tough, power-packed machine for the heaviest work. And a fast, economical one for the light jobs which take up most of your working time.

No doubt about it, the low-priced Ferguson is your best buy in tractors today — whether you work a smallholding or hundreds of acres.

Ask your Ferguson Dealer for a free demonstration on your own farm. Ask him, too, about the Ferguson Pay-as-you-Farm Plan.

SEE FERGUSON FIRST

Ferguson tractors, sold by Massey-Harris-Ferguson (Sales) Ltd., are manufactured by The Standard Motor Co. Ltd., Coventry

Very clean ploughing with three furrows uphill. December 1954.

February, 1954 FARM MECHANIZATION 27

The Ferguson ploughing three furrows up a 1 in 3½ hill.

PLOUGHING UP STEEP HILLS

It is a fallacy that bigger, heavier wheeled tractors can beat the Ferguson ploughing up steep hills.

Heavy wheeled tractors need most of their power to propel their great weight uphill, leaving little to spare for work. Their implements add still more wasteful weight.

Not so with the Ferguson.

Because of its unique design, the Ferguson eliminates built-in weight, transfers the natural forces of the implement in work to add penetration and traction. This gives perfect performance under all conditions, saves fuel and provides power to spare for the steepest hills.

Prove it for yourself by asking the salesman who claims his tractor is the equal of the Ferguson, to meet the Ferguson in open competition on the steepest hills. Also, insist that a full competitive demonstration be made with a full range of equipment. This will prove beyond doubt that the Ferguson meets more of the needs of all of the farmers, for more jobs, more economically, than any other tractor.

MASSEY-HARRIS-FERGUSON

Ferguson tractors are manufactured for Massey-Harris-Ferguson, Coventry, by The Standard Motor Company Ltd.

Complete mastery of the slopes with Ferguson tractor and plough. Note the company name change to Massey-Harris-Ferguson after Massey-Harris purchased the Ferguson company. 1954.

Even the world's premier cricket ground could not manage without Ferguson!

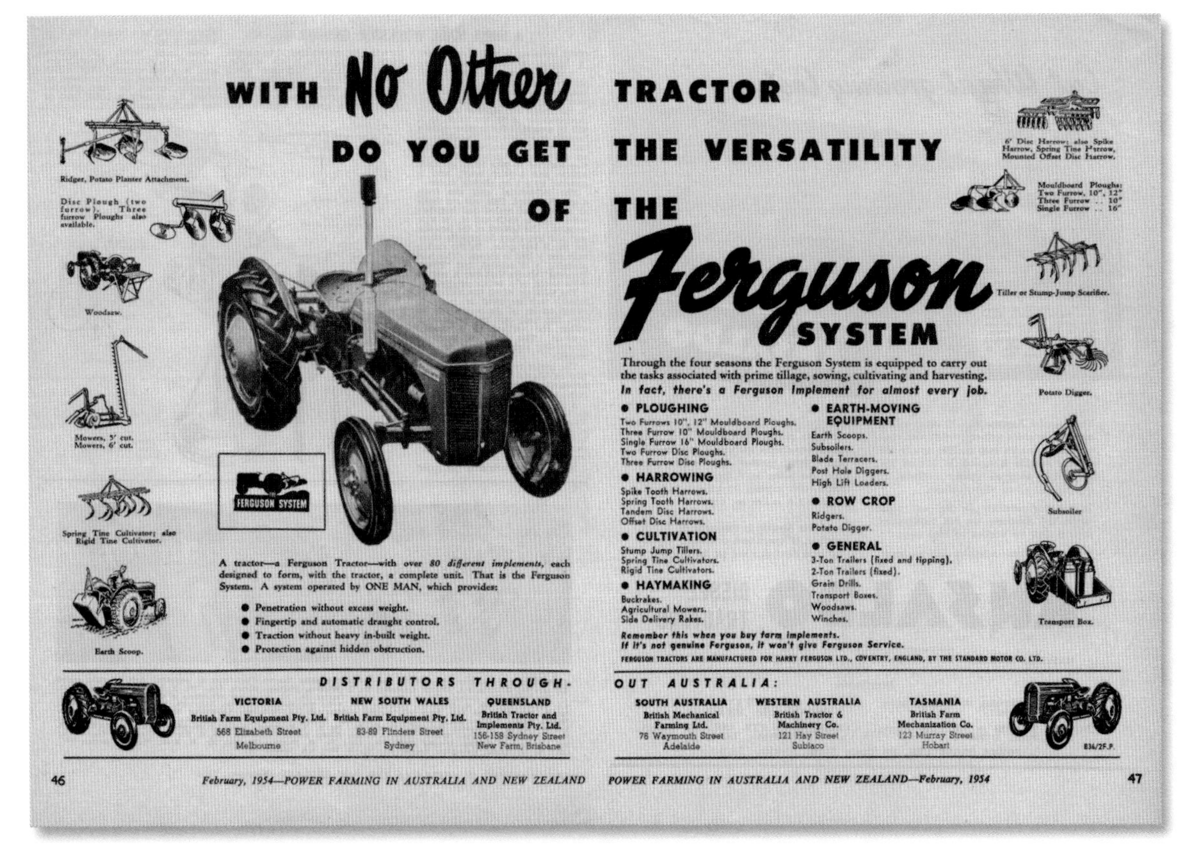

By 1954 there were 80 different implements available for the Ferguson tractor giving it astonishing versatility. February 1954.

This farmer can't get on to his land
This farmer has a Ferguson*

★ *Ferguson's exclusive weight transfer feature is the secret! It enables Ferguson owners to work soft, wet ground and farm steep slopes with confidence — and without harmful packing of the soil!*

FERGUSON ***the world's most versatile tractor***

does more, saves more, all the year round . . .

THE FARMER who owns a Ferguson can recover from the effects of waterlogged land much more quickly than his neighbour with other tractors! Ferguson does not need costly excess weight to gain traction and pull . . . 'fuel-stealing' dead weight has been eliminated. That's why the low-weight Ferguson, with Ferguson System implements, enables owners to be first in the field, and first out!

All round-the-year, all round the farm, it costs less to farm the Ferguson way. See for yourself! Ask your Ferguson Dealer for a free demonstration on your farm.

Ferguson outvalues them all at a price you can afford to pay.

Go years ahead — Go Ferguson!

ii October, 1955—POWER FARMING IN AUSTRALIA AND NEW ZEALAND

WHATEVER THE JOB — THERE'S A FERGUSON TRACTOR-MATCHED IMPLEMENT TO DO IT PERFECTLY!

FERGUSON MOWERS designed for fast cutting in all types of hay. Cut square corners without stopping.

SIDE DELIVERY RAKE. Fast raking up to 7 acres per hour. True 'sideward' raking action gets all the hay.

FERGUSON 2-TON TRAILER. Designed for safety on hillsides . . . 30 cwt., 3-ton, fixed and tipping trailers also available.

FERGUSON BUCKRAKE. Engineered for easy handling and fast transport. Width of tines 8′ 3″.

FERGUSON SYSTEM

● ***Ferguson Tractors and Implements are distributed throughout Australia by . . .***

VICTORIA
BRITISH FARM EQUIPMENT PTY. LTD.
568-576 ELIZABETH ST., MELB. FJ 1181

QUEENSLAND
BRITISH TRACTOR & IMPLEMENTS PTY. LTD.
156-158 SYDNEY ST., NEW FARM. LW 1011

WEST AUSTRALIA
BRITISH TRACTOR & MACHINERY CO.
121 HAY STREET, SUBIACO, W.A. W 2626

NEW SOUTH WALES
BRITISH FARM EQUIPMENT PTY. LTD.
602 BOTANY RD., ALEXANDRIA. MU 3901

STH. AUSTRALIA
BRITISH MECHANICAL FARMING LTD.
76 WAYMOUTH ST., ADELAIDE, S.A. LA1787

TASMANIA
BRITISH FARM MECHANISATION CO.
123 MURRAY ST., HOBART, TAS. B 2861

FERGUSON DEALERS, SERVICE AND SPARE PARTS EVERYWHERE THROUGHOUT AUSTRALIA

FS/2033

POWER FARMING IN AUSTRALIA AND NEW ZEALAND—October, 1955 iii

When the going gets rough the Ferguson keeps going. October 1955.

JUST ONE LEVER – GIVES YOU FULL CONTROL !

when you farm with FERGUSON

One lever at your fingertips — the Ferguson System of farming is as simple as that. Just set your finger-tip lever where you want it and away you go.

You are relieved of the need for constantly watching implements, checking soil conditions and wondering about the quality of work being done. The hydraulic system gives control of the implement in the soil, maintains proper traction on uneven ground and makes it possible for you to farm steep hills, wet and boggy ground where no other tractor can go.

Ferguson provides the 'missing link' between tractor and implement — an exclusive hydraulic system controlled by one single lever that helps you with all your farm tasks — from ploughing to sawing wood . . . digging ditches to grading roads . . . pulling stumps to digging post holes . . . mowing hay to farm transport, and all the other odd jobs that make up the farming year.

ONLY FERGUSON GIVES YOU . . ALL THESE ADVANTAGES !

- Penetration without excess weight.
- Fingertip & Automatic Draft Control.
- Traction without built-in weight.
- Automatic protection against hidden obstructions.
- Keeps the front-end down.
- Quick raising & lowering of implements.

PETROL, KERO, & DIESEL MODELS

YOU ARE YEARS AHEAD WHEN YOU **FARM WITH FERGUSON**

● *Ferguson Tractors and Implements are distributed throughout Australia by . . .*

VICTORIA
BRITISH FARM EQUIPMENT PTY. LTD.
568-576 ELIZABETH ST., MELB. FJ 1181

QUEENSLAND
BRITISH TRACTOR & IMPLEMENTS PTY. LTD.
156-158 SYDNEY ST., NEW FARM. LW 1011

WEST AUSTRALIA
BRITISH TRACTOR & MACHINERY CO.
121 HAY STREET, SUBIACO, W.A. W 2626

NEW SOUTH WALES
BRITISH FARM EQUIPMENT PTY. LTD.
602 BOTANY RD., ALEXANDRIA. MU 3901

STH. AUSTRALIA
BRITISH MECHANICAL FARMING LTD.
76 WAYMOUTH ST., ADELAIDE, S.A. LA1787

TASMANIA
BRITISH FARM MECHANISATION CO.
123 MURRAY ST., HOBART, TAS. B 2861

FERGUSON DEALERS, SERVICE AND SPARE PARTS EVERYWHERE THROUGHOUT AUSTRALIA

FS/2036

122 *September, 1955—POWER FARMING IN AUSTRALIA AND NEW ZEALAND*

POWER FARMING IN AUSTRALIA AND NEW ZEALAND—September, 1955 123

It's amazing what can be put under the control of one hand with only a light fingered touch. Effortless! September 1955.

108 FARMER & STOCK-BREEDER, 18-19 January 1955

Judge the Ferguson multi-purpose Seed Drill and Fertilizer Attachment for yourself!

One drill to do the work of four implements! With this Ferguson Grain & Fertilizer drilling team, all types of seed and fertilizer are sown and distributed fast and accurately — by one man. But first, look at these comparative costs:

PURCHASE PRICE

Ferguson Seed Drill and Fertilizer Attachment.	4 implements required to do same work at the following conservative costs:	
	Grain drill	£140
	Root drill	£50
	Grass barrow	£20
	Fertilizer distributor	£50
Purchase price: £229.10	Total purchase price:	£260

WORKING COSTS

Ferguson Seed Drill & Fertilizer attachment working 16 acres at 2½ acres per hour:		Conventional drill working 16 acres at 2½ acres per hour:	
1 man	16.3	2 men	1.12.6
tractor plus depreciation	1.3.7	tractor plus depreciation	1.3.7
drill plus depreciation	19.2	drill plus depreciation	19.2
Total	£2.19.0	Total	£3.15.3

Saving 12% on purchase

Saving 21% on working costs

plus saving of storage space for 4 implements.

You get value for your money, too!

Seed Drill: Waterproof 6 bushel hopper. Coulters easily spaced, easily detachable. Single speed drive. Retractable jack for one man hitching up. Seed-flow automatically cut-off when coulters raised out of work. Sows wide range of seed from clover to maize.

Fertilizer Attachment: Separate feed unit mechanism for fertilizer. Uniform flow of fertilizer at rates from 110 to 1,200 lbs. per acre. All gears and bearings self-lubricating. Ask your Ferguson Dealer for a free demonstration on your farm, and about the Ferguson Pay-as-you-Farm Plan.

SEE Ferguson FIRST!

Ferguson tractors, sold by Massey-Harris-Ferguson (Sales) Ltd., are manufactured by The Standard Motor Company Ltd., Coventry.

Cheaper to buy and cheaper to operate. Multi-purpose versatility. January 1953.

July, 1955 FARM MECHANIZATION 27

STOP WORRYING about breakdowns — start farming with Ferguson! Ferguson tractors and implements are made of the finest materials. Made to go on *and on* working. And they are backed by Ferguson Service. That means regular inspection and maintenance by *experts*, ensuring that Ferguson equipment is *always* in peak condition.

Find out for yourself what Ferguson can do! Ask your Ferguson Dealer for a free demonstration on your own farm.

QUALITY PAYS—FARM WITH FERGUSON

Ferguson tractors, sold by Massey-Harris-Ferguson (Sales) Ltd., are manufactured by The Standard Motor Co. Ltd., Coventry.

D3

That poor farmer next door! What implement is carrying that stack of hay? Is it the elusive but much talked about Ferguson rick lifter? The side member of the implement looks too high for the normal Ferguson Buckrake. July 1955.

May, 1955 FARM MECHANIZATION 25

This farmer can't plough his hill

This farmer has a Ferguson

THE FERGUSON SYSTEM gives greater stability and traction without excess weight. It has ploughed hills impossible for other tractors.

See for yourself! Ask your Ferguson Dealer for a demonstration on your own farm—any time, any job.

QUALITY PAYS—FARM WITH FERGUSON

Ferguson tractors, sold by Massey-Harris-Ferguson (Sales) Ltd., are manufactured by The Standard Motor Company Limited, Coventry

D

As well as ploughing uphill, the Ferguson could plough across slopes. May 1955.

Oct. 1, 1955 FARM IMPLEMENT AND MACHINERY REVIEW 1031

This farmer is only partly mechanised

This farmer has Ferguson

WHAT'S THE GOOD of buying a tractor—without using it for every job on the farm? You waste man-hours on work machinery could do in a quarter the time; you leave expensive machinery lying idle. Ferguson farmers keep their tractors working, even digging silage pits! For there's a complete range of equipment in the Ferguson System — an implement for every job at every season. Each job is done quickly, on little fuel, by one man. That's how Ferguson saves time, saves labour, cuts costs. That's why it pays to be completely mechanised — the Ferguson way.

See for yourself! Test the Ferguson range where you're going to use it — on your own farm. Your Dealer will give you a free demonstration.

QUALITY PAYS—FARM WITH FERGUSON

Ferguson tractors, sold by Massey-Harris-Ferguson (Sales) Ltd., and manufactured by The Standard Motor Company Ltd., Coventry.

E

The farmer next door continues to suffer! He'll get the Ferguson message when his hands get sore and his back starts to ache. October 1955.

A farmer selling the Ferguson System to his neighbour. January 1956.

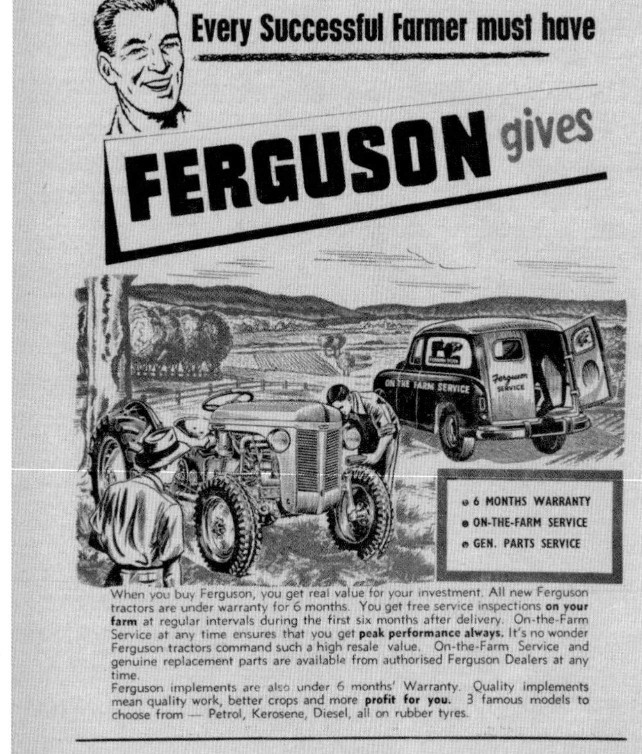

Look in the small print and you will see that Ferguson tractors always command a high resale value. July 1956

POWER FARMER, *April 1956*

500,000 Ferguson

tractors made and sold since 1946

No farmer buys a tractor without thinking hard and long. He has a wide choice of makes and models; he's going to use his tractor for a long time; and he's going to use it for a thousand different jobs.

That's why we're proud that 500,000 Fergusons — the same model, all the time — have been made and sold since 1946. 500,000 times a farmer has *thought*, and made the decision — "the British Ferguson is the best tractor for my money". And they know that decision was right — 500,000 times!

What it means to the country

In the short time since production commenced in 1946, Ferguson products have earned Sterling Exports of £120,000,000, and for the peace of the world have created more contentment in over 117 countries by producing more food at less cost.

"There is no better tractor"

We take this opportunity of expressing our appreciation to The Standard Motor Co. Ltd., and all their supporting Suppliers, without whose effort this great achievement would not have been possible.

339

Made it at last. Half a million tractors made in the UK. Sadly Harry Ferguson was no longer at the helm or even owned the company. But his name proudly lived on. By now 117 countries are benefiting form the Ferguson System. April 1956.

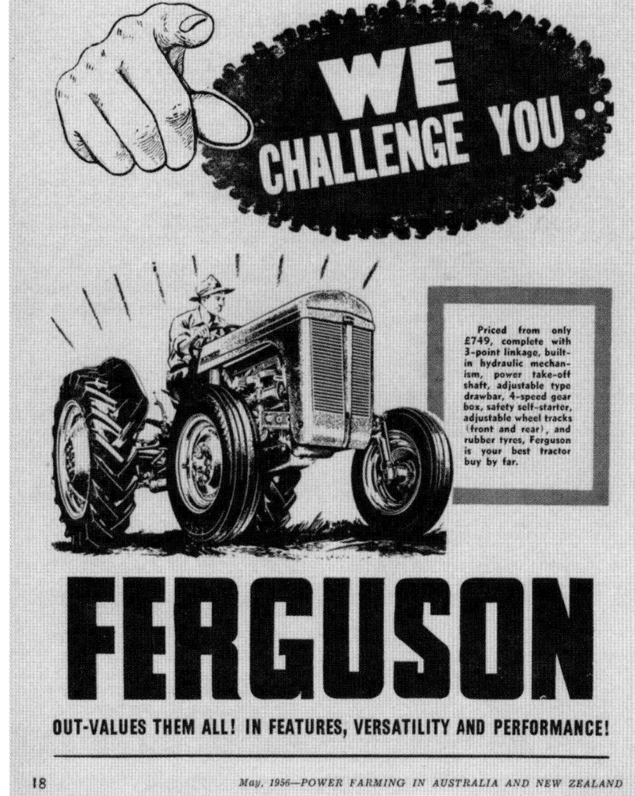

Nearing the end of the production period of the TE Ferguson tractors now and they are seen to have about doubled in price since their introduction. May 1956.

NOW! reconditioned Fergusons

...with a GUARANTEE

The symbol shown here is *news* for all you who've long wanted — and needed — a Ferguson. It stands for the Ferguson Reconditioned Tractor Scheme ; and it's your guarantee that the Ferguson you buy has been *fully* reconditioned by Ferguson-trained experts.

When you buy a used Ferguson under the Ferguson R.T.S. you get the Dealer's two-month two-star guarantee or four-month four-star guarantee. And you get that built-in quality which makes a Ferguson go on *and on* working — and keeping its value. For all this you pay a price well within your means — and payment facilities similar to the Ferguson Pay-as-you-Farm Plan are available.

Your Ferguson Dealer will give you full details, and specifications, of the Ferguson Reconditioned Tractor Scheme. See him today!

FERGUSON – first again!

Ferguson tractors, sold by Massey-Harris-Ferguson (Sales) Ltd., are manufactured by The Standard Motor Co. Ltd., Coventry

152 POWER FARMER, *March, 1956*

The tractor reconditioning service gave you back a tractor with a red engine. This is one of the first advertisements to clearly show a diesel engine tractor. March 1956.

April 1, 1956 FARM IMPLEMENT AND MACHINERY REVIEW 2075

The Ferguson System

and its imitators

A straightforward statement of facts

Ever since Ferguson revolutionised farm mechanisation, tractor manufacturers have flattered the System by trying to imitate it : but don't imagine their devices make their tractors as good as the Ferguson. Ferguson engineers tried and discarded most of these ideas more than twenty years ago. Now read why only " the world's most copied tractor " gives you the most efficient method of weight transference and, hence, genuine automatic depth control and protection against hidden obstacles.

Wheel grip and weight transference

A tractor is useless unless it grips the ground. It can do this in one of two ways. Through massive weight built into itself — which wastes power and fuel, and makes the tractor more expensive to buy. Or through weight transference — harnessing the weight and force of the implement working in the soil.

In some tractors weight transference is achieved by fixed linkage : all very well when the land is dead level and soil 100 per cent uniform — but how often do you find these conditions ?

In other tractors, weight transference comes from various devices in the hydraulic system : depth control wheels, or other forms of depth gauge, are necessary ; only a small part of the implement's weight can be transferred and, when weight transference nears 50 per cent, the implement's response to different soil conditions is too sluggish to be of real benefit in difficult situations.

Only in the Ferguson are the weight and force of the implement transferred to the tractor wheels in the simplest and most natural way possible : through the tractor's linkage and a hydraulics mechanism which is *built-in* — not just added as an afterthought.

An automatic " nerve centre "

This exclusive Ferguson hydraulics mechanism is the " nerve centre " of the tractor and implement. It *automatically* corrects excessive or insufficient stress in either of them almost instantly — and so gives you two all-important benefits :

1. Whatever conditions of soil your implement meets, your tractor *automatically* adjusts its weight to the job, and the implement keeps working at an even depth. No need for depth control wheels. No need for the operator to make hurried adjustments.

2. When your implement meets a hidden rock or root, weight transference is *automatically* eliminated, the tractor *automatically* stops, and its front wheels are *automatically* forced down. No danger for your implement — or operator.

Prove it for yourself

Look around. Ask your friends. Study the catalogues. Read the advertisements. You'll see for yourself that no imitation can equal the Ferguson. Then ask your dealer for a free demonstration — of " the world's most copied tractor " !

Ferguson

" IMITATED—BUT NOT EQUALLED "

Ferguson tractors, sold by Massey-Harris-Ferguson (Sales) Ltd., are manufactured by The Standard Motor Co. Ltd., Coventry.

By the mid 1950s Ferguson tractors had the competition in a state of frustration. They had hydraulics to offer, but not the key features of the Ferguson System's depth control, weight transfer and safety protection. However the intensity of the opposition's claims seems to have led to M-H-F needing to explain and re-assert the superiority of the Ferguson System of hydraulic implement control. April 1956.

A much used photo. Now the real horses are all but gone from British farms, the Ferguson tractor becomes classed as a "willing horse". May 1956.

Bring in the experts when you want to prove a point. Here M-H-F quotes National Institute of Agricultural Engineering tests on the Ferguson. July 1956.

82 The Farmers Weekly, September 28, 1956

NEW! FERGUSON SPINNER BROADCASTER

Now! Spread fertiliser faster, more cheaply!

One-man operation with the new Ferguson mounted Spinner Broadcaster gives you a fast economical way of spreading fertiliser, lime, seeds, salt and grit. Through close coupling on the tractor linkage, the Spinner Broadcaster withstands high speeds over rough ground . . . spreads fertiliser to a width of up to 50 feet.

Mounted Spinner Broadcaster:

★ easily attached and detached ★ has large capacity hopper – holding $4\frac{1}{2}$ cwt. of nitro chalk, $7\frac{1}{2}$ cwt. ground limestone ★ has easily-adjusted sowing rates, from $\frac{1}{2}$ cwt. to over 2 tons per acre ★ is easily cleaned in a matter of minutes ★ has slow-speed agitator revolving in opposite direction to spinner – breaking up lumps, preventing compaction of fertilisers, giving an even feed.

See your FERGUSON Dealer now!

FERGUSON TRACTORS, SOLD BY MASSEY-HARRIS-FERGUSON (SALES) LTD., ARE MANUFACTURED BY THE STANDARD MOTOR CO. LTD., COVENTRY

Yet another new implement is added to the list in the last year of production of the famous "grey Fergie" TE20 tractor. September 1956.

Australian COUNTRY, February, 1957 83
EVERY SUCCESSFUL FARMER MUST HAVE
Complete after-sales service!
FERGUSON
gives you
SERVICE SECOND TO NONE
ON THE FARM SERVICE
Ferguson SERVICE
• 6 MONTHS WARRANTY
• ON-THE-FARM SERVICE
• GENUINE PARTS SERVICE
When you buy Ferguson, you get real value for your investment. All new Ferguson tractors are under warranty for 6 months. You get free service inspections on your own farm at regular intervals during the first six months after delivery. On-the-Farm Service at any time ensures that you get peak performance always. It's no wonder Ferguson tractors command such a high resale value. On-the-Farm Service and genuine replacement parts are available from authorised Ferguson Dealers at any time.
Ferguson implements are also under 6 months' Warranty. Quality implements mean quality work, better crops and more profit to you.
Get the full facts today by mailing coupon below. 3 famous models to choose from—Petrol, Kerosene, Diesel, all on rubber tyres.
Before you buy—think . . .
WHAT "ON-THE-FARM SERVICE" MEANS TO YOU
When you buy Ferguson, you get real value for your money. Not only do you get a true working partner — one that will help you every day of every month of every year to make your task easier—but in addition you get after sales service that is second to none—service that safeguards your investment and keeps the value in your tractor.
Over 54,000 Australian farmers have proved Ferguson to be their greatest farming investment. Ferguson is your best investment, too, because it offers more value at a price you can afford to pay. Ferguson has exclusive features not found in any other tractor — features you must have to do better work with less effort and at lower cost.
you must have to do better work with less too — Ferguson owners get it — it's just another way of of making sure you get full value for your purchase.
Get the Facts!
MAIL THIS COUPON NOW!
BRITISH FARM EQUIPMENT PTY. LTD.
Please send me your free Booklet "Ferguson After-Sales Service".
NAME
ADDRESS
ACM2
BRITISH FARM EQUIPMENT PTY. LTD.
A MEMBER COMPANY OF THE STANDARD MOTOR PRODUCTS GROUP
NEW SOUTH WALES
602-612 BOTANY RD., ALEXANDRIA
TELEGRAMS: "BRITFARM." PHONE: MU 4021
FERGUSON SYSTEM
VICTORIA
568-576 ELIZABETH ST., MELBOURNE
TELEGRAMS: "BRITFARM." PHONE: FJ 0221

By this date 54,000 Australian farmers are reported as having bought a Ferguson tractor. February 1957.

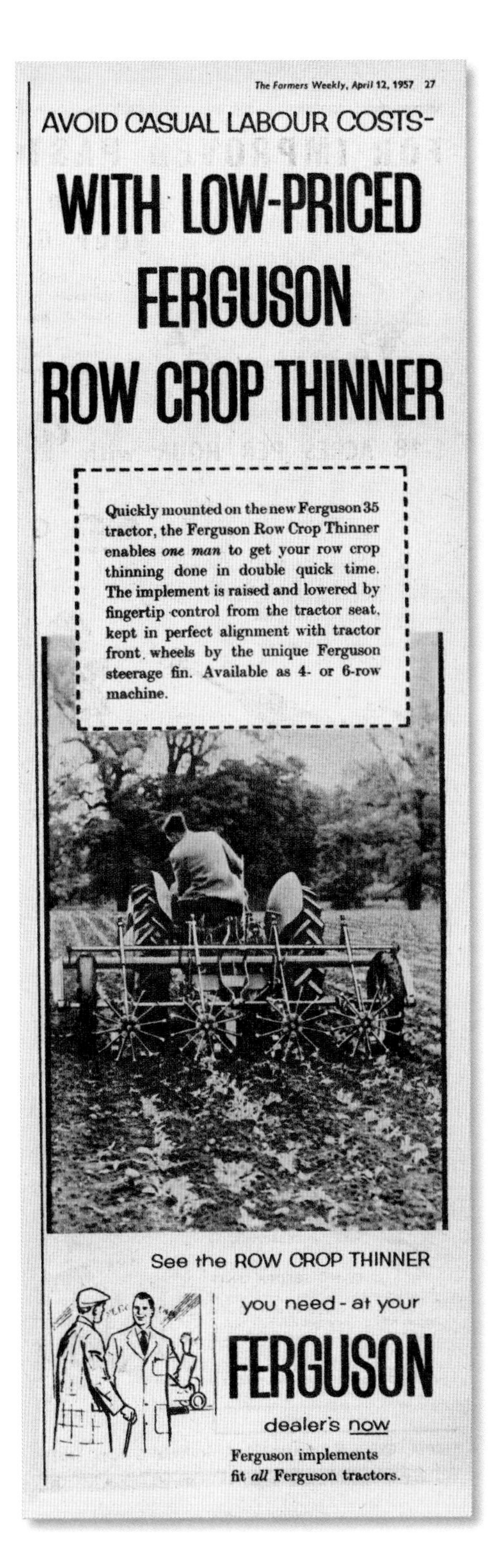

The Farmers Weekly, April 12, 1957 27
AVOID CASUAL LABOUR COSTS-
WITH LOW-PRICED FERGUSON ROW CROP THINNER
Quickly mounted on the new Ferguson 35 tractor, the Ferguson Row Crop Thinner enables *one man* to get your row crop thinning done in double quick time. The implement is raised and lowered by fingertip control from the tractor seat, kept in perfect alignment with tractor front wheels by the unique Ferguson steerage fin. Available as 4- or 6-row machine.
See the ROW CROP THINNER
you need - at your
FERGUSON
dealer's now
Ferguson implements fit *all* Ferguson tractors.

Advertising of the TE20 tractor continued into 1957 to clear old stocks alongside the new "grey-gold" Ferguson 35 which had been introduced in 1956. A useful but now relatively rare implement for specialist work is the row crop thinner. April 1957.

The Farmers Weekly, April 12, 1957 25
HOE EVERY INCH OF EVERY ROW
WITH A FERGUSON STEERAGE HOE
Simply mounted on the new Ferguson 35 tractor. Ferguson Steerage Hoes get your hoeing done with great accuracy, and in less time than ever before. Easy row width adjustment from 16" to 24". Two models available—Rigid Type and Independent Gang Type.
See the STEERAGE HOE
you need - at your
FERGUSON
dealer's now
Ferguson implements fit *all* Ferguson tractors

Another specialist row crop machine is the steerage hoe which did need another man besides the driver. April 1957.

The Farmers Weekly, April 26, 1957 23
COLLECTING, CARRYING, DUMPING SILAGE-
ONE MAN'S WORK WITH FERGUSON BUCKRAKE
Simply mounted on the new Ferguson 35 tractor, the Ferguson Buckrake is raised, lowered and tipped from the tractor seat. Light, yet strong, the Buckrake collects, carries and dumps loads up to 750 lbs. Exclusive Ferguson hydraulics prevent tines digging in.
See the BUCKRAKE
you need - at your
FERGUSON
dealer's now
Ferguson implements fit *all* Ferguson tractors.

The buckrake was a widely used and abused simple implement with no moving parts. Frequently overloaded, frequently driven into loads incorrectly, frequently bringing the front of the tractor into the air and causing the driver to use independent brakes to steer! April 1957.

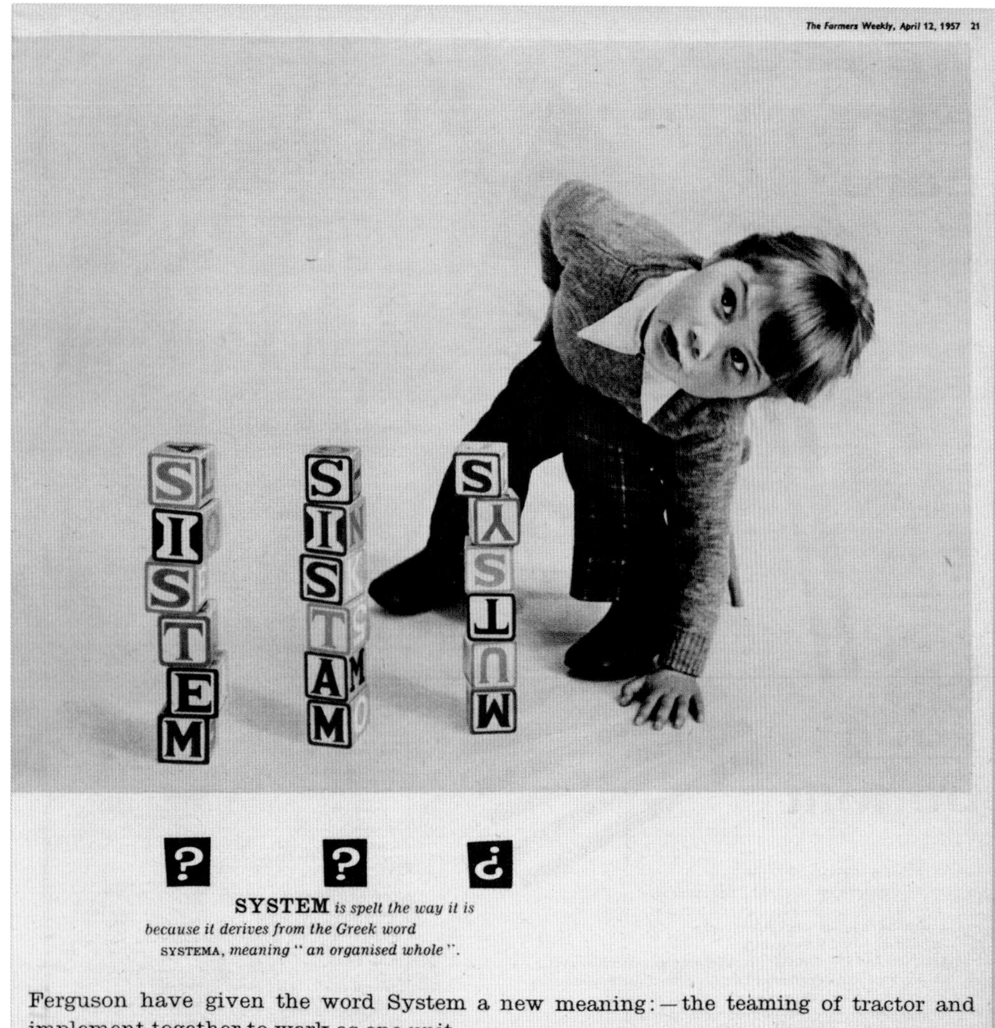

? ? ¿

SYSTEM *is spelt the way it is because it derives from the Greek word* SYSTEMA, *meaning " an organised whole ".*

Ferguson have given the word System a new meaning: — the teaming of tractor and implement together to work as one unit.

This new meaning is important to you, as a farmer, because the principle of teamwork is the core of the Ferguson System. This principle, embodied in the basic design of all Ferguson equipment, enables you to work faster and more economically with a Ferguson tractor and implement than with any other farm machinery; and thus to produce more food at less cost. That is why it is well worth your while to study the Ferguson equipment on the following pages.

The Ferguson System means teamwork in farm machinery
means more food at less cost

Child's play. Ferguson made mechanised farming a team effort between tractor and implement. A new meaning for the concept of a system. April 1957.

27

CUT KALE-CUTTING COSTS!

Now you can cut and carry kale in one go — with the Ferguson Tractor and Kale Cutrake. It's a one-man job, four times faster than by hand!

By reversing into the crop the Cutrake cuts and loads at the same time. You get a cleaner crop. You leave a clean, level stubble. In about 12 seconds you can load between 3 cwt. and 5 cwt., according to the crop. Kale cutting is changed from a misery to a quick and pleasant job, with the Ferguson Kale Cutrake.

You save time, labour, money. That's Ferguson farm mechanisation for you!

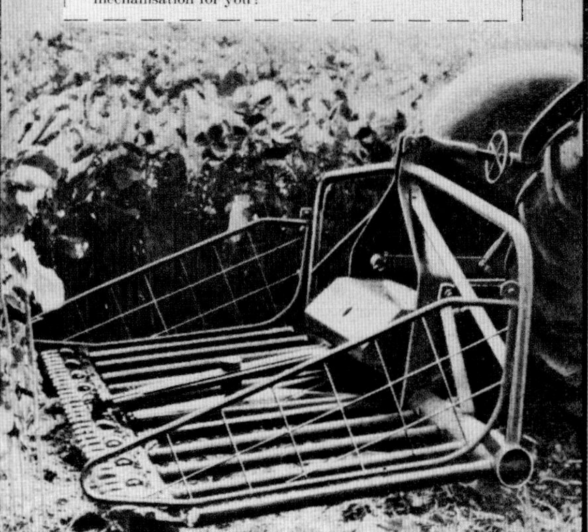

See the Kale Cutrake at your Ferguson dealer's now

STANDARDISE with FERGUSON

The kale cutter saved many hours of arduous and often wet, unpleasant work for livestock farmers. April 1957

The Farmers Weekly, July 19, 1957 25

ANSWERS EVERY HEDGECUTTING PROBLEM

At your Ferguson Dealer's, now, there's the implement you need to make light work of your most awkward hedges and ditches. With a Ferguson Hedgecutter, 60 ft. of pressure hose enables the cutter bar to work on grass banks, in awkward ditches, cut water-bound weeds, both sides of the hedge — without moving your Ferguson Tractor . . . gets the work done far faster and more efficiently than ever before!

HEDGECUTTER UNIT • is made from light alloys and weighs only 8¼ lbs. • has special saw section to deal with all timber up to 6 inches diameter • 37″ or 39″ cutter bar lengths supplied • resists damage from hidden wire and obstructions • also valuable for pruning fruit trees and clearing scrub.

COMPRESSOR UNIT • of proved sliding vane type • quickly and easily mounted on the tractor • also useful for paint spraying, disinfecting, working air tools, etc.

The Ferguson Hedgecutter is manufactured by R. M. Marples and Son Ltd., sole manufacturers and patentees.

STANDARDISE with FERGUSON

The hedge cutter was jointly badged Ferguson and Marples. It was astonishingly light to use because it employed much aluminium in its construction. July 1957.

The zenith of the "grey Fergies'" career was undoubtedly when they reached the south pole having crossed the Antarctic. Remember they used Zenith carburettors! Massey-Ferguson (now superseding Massey-Harris-Ferguson) made strong use of the Ferguson reputation in this advertising over a year after production of the tractor had ceased. January 1958.

THE FERGUSON TO20

The curtain rises on a new era for Ferguson in the USA. Harry Ferguson produces his own tractor after the break with Ford. Welcome to the Ferguson TO20. This is possibly one of the first adverts for the tractor. April 1948.

The Ferguson TO20 was in reality the USA version of the TE20. With the cessation of Ford Ferguson tractor production in 1947 in the USA, Harry Ferguson needed to urgently resurrect production of a Ferguson tractor there. Although he never wanted to be a manufacturer his hand was in effect forced and he built a totally new plant in Detroit. The TO20 was fitted with the same Continental Motors engine as the first Ferguson TE tractors. Otherwise it was essentially the same as the TE tractor. Early in the production of the TO tractors parts were in fact shipped across from the UK. The first TO came off the production line in October 1948. Remarkably the factory had been built and commissioned in about eight months. Unlike for the TE tractors there were essentially no model variants.

Beginning . . . another great era in power farming!

A GREAT NEW FARMING MACHINE to help you cut the costs of food production still more! A powerful, rugged tractor with many new high-performance features designed from the experience of over 300,000 Ferguson System users. Stepped-up power lets you walk away with two 14-inch plows—with reserve for the tough spots. You can brake both wheels with a single pedal, or use the individual brake pedals for turning. And many other improvements for greater safety, economy and convenience.

A MISER ON FUEL! Powerful Continental valve-in-head engine, combined with the weight-saving and automatic traction control of the Ferguson System, gives you more work from a gallon of fuel than you ever dreamed possible! High-torque design to give you amazing lugging power at low engine speeds. Responsive governor, fully enclosed from trash and dust. High-efficiency cooling fan and removable "wet" sleeves with full-length water circulation. An engine built to "take it!"

A NEW SYSTEM FOR LOW-COST HAULING with revolutionary method of linking trailer to tractor. With this new Ferguson System of transportation you can haul loads of 3 tons or more out of soft fields, up steep grades, on slippery roads or over rough ground. On countless farms, located at average local market distances, this latest Ferguson invention slashes costs of hauling farm products—a revolutionary new solution to your farm transportation problems.

ONLY THE FERGUSON SYSTEM, invented, perfected and patented by Harry Ferguson, and authorized *only* for use in the new Ferguson Tractor, gives you *all five* of these big advantages: **1.** Penetration without excessive implement weight. **2.** Both finger tip and automatic hydraulic depth control. **3.** Traction without excessive built-in weight. **4.** Keeps the tractor's front end down. **5.** Automatically protects both implement and tractor when the implement strikes an obstruction.

FERGUSON TRACTOR AND FERGUSON SYSTEM IMPLEMENTS

See it at your authorized Ferguson Tractor dealer

Harry Ferguson, Inc., 3639 E. Milwaukee, Detroit 11, Mich.

Only The FERGUSON SYSTEM gives you
ALL THESE BIG ADVANTAGES!
1 PENETRATION WITHOUT EXCESSIVE IMPLEMENT WEIGHT. Action of the Ferguson System pulls the implement into the ground. Costly weight . . . as much as several hundred pounds for some implements—is eliminated.
2 FINGER TIP HYDRAULIC DEPTH CONTROL. Ferguson hydraulic implement control is both manual and automatic. This hydraulic control is not a separate attachment. It is a built-in, integral part of the exclusive Ferguson System. It costs you nothing extra.
3 TRACTION WITHOUT EXCESSIVE BUILT-IN WEIGHT. The Ferguson System uses forces that result from pulling an implement through the soil to adjust traction as needed, automatically, for either light or heavy work. The saving of hundreds of pounds in tractor weight saves fuel every trip through the field.
4 KEEPS THE TRACTOR'S FRONT END DOWN. As a Ferguson unit implement is pulled forward in the soil, the top link of the Ferguson 3-point linkage exerts a strong forward and downward force. This keeps the front wheels down and provides positive, safe steering control.
5 AUTOMATICALLY PROTECTS TRACTOR AND IMPLEMENT WHEN IMPLEMENT STRIKES AN OBSTRUCTION. No time lost resetting separate protective devices.
GREAT STRENGTH is combined with weight saving in Ferguson Implements due to the Ferguson System and use of special alloy steels new to the implement industry.
YOUR WORK IS MADE EASIER than ever before, because only the Ferguson System combines the muscle-saving advantages of both (1) finger tip and (2) automatic hydraulic control.
ASTONISHING FUEL ECONOMY results from the action of the Ferguson System in providing traction without excessive built-in weight.
It's THE FERGUSON SYSTEM
THAT MAKES THE DIFFERENCE!
HUGE NEW FACTORY FOR THE FERGUSON SYSTEM
Purchased by Harry Ferguson, Inc., this huge plant at Cleveland, Ohio, will soon be one of the nation's largest, most modern tractor factories. More than ever, your Ferguson Dealer will be in position to supply and service mechanized farm equipment embodying the exclusive Ferguson System.
HARRY FERGUSON, INC.
© 1947, Harry Ferguson, Inc.
DETROIT 3, MICHIGAN
38
SUCCESSFUL FARMING, MAY, 1947

Announcing the Ferguson System and new factory for production of Ferguson tractors. Strangely it states that the factory was purchased in Cleveland, Ohio whereas the factory was actually built at Detroit. May 1947.

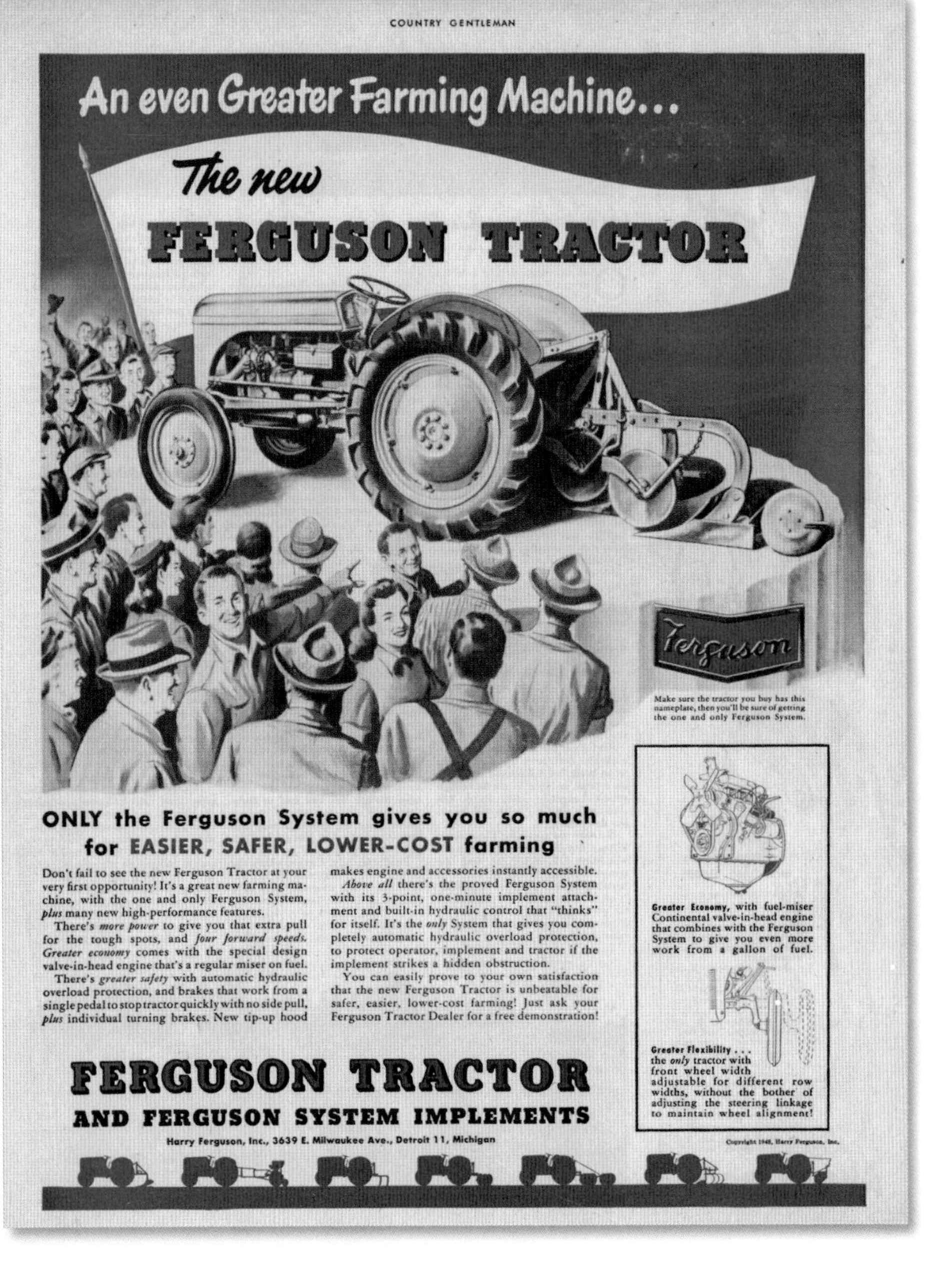

The crowds come out to see the new Ferguson tractor – even ladies came out to view it! June 1948.

Looks as if they are coming straight down the production line and into action! Better from every angle? It does look more refined than its predecessor produced in the USA, the Ford Ferguson. August 1948.

Make use of the Ferguson PTO with a Belle City corn picker – an approved Ferguson implement. Again that theme – one man for one job. October 1948.

76 FARM JOURNAL

Measure TRACTION..

YOU'LL CHOOSE THE NEW **FERGUSON TRACTOR!**

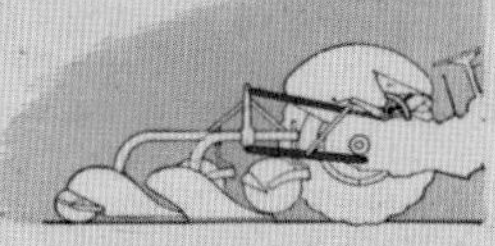

Your next tractor *could* cost you hundreds of wasted dollars over the years in needless upkeep and fuel bills on a big, expensive machine—all because of the often mistaken idea that a lot of built-in weight is the only way to make sure of ample traction for the tough pulling.

With the Ferguson System, there's no need for ballast weight. *Traction is developed automatically as needed* for each different job. 3-Point Linkage of implement to tractor and hydraulic control make use of natural forces to automatically *increase* weight—and traction—to suit the job.

Traction—created by the one and only Ferguson System—gives the Ferguson Tractor power efficiency equal to tractors of much larger size and greater weight.

The specially-designed Continental valve-in-head engine stresses economy as well as more than ample 2-plow power. It *saves* on fuel all through the long hours of heavy running.

Before you buy a new tractor, measure the crop-to-crop and job-to-job performance of the New Ferguson against *any* other tractor. You, too, will choose the New Ferguson with the one and only Ferguson System. Make a date now with your friendly Ferguson Dealer for *your* demonstration.

THE FERGUSON SYSTEM CHANGED THE TREND OF TRACTOR AND IMPLEMENT DESIGN

The Ferguson System uses a different basic principle of applying and controlling power . . . with greater efficiency, greater safety. Implement is attached to tractor with two bottom links which *pull*, and a top link which *pushes* forward and downward above the rear axle. Tractor and implement become a single, self-propelled unit. The built-in hydraulic system provides *both* Finger Tip and Automatic Implement Control, and the exclusive automatic overload protection.

FOR A BETTER LIVING AND A BETTER WORLD THROUGH . . .
Lower production costs and increased profits for the farmer . . . Lower food costs for the consumer . . . Less world unrest from hunger and want . . . Greater security for world peace.

FERGUSON TRACTOR AND FERGUSON SYSTEM IMPLEMENTS

For the one and only Ferguson System, look for this nameplate.

Harry Ferguson, Inc., 3639 E. Milwaukee Ave., Detroit 11, Mich.

Copyright 1949, by Harry Ferguson, Inc.

That tiller is certainly in to some depth, and no sign of the tractor skidding with increased traction provided by weight transfer. March 1949.

APRIL, 1949 29

FOR THE WIDEST RANGE OF FARM JOBS

FERGUSON SYSTEM MAKES THE DIFFERENCE

Model TO-20
Ferguson Tractor Illustrated

Measure JOB-FLEXIBILITY...

YOU'LL CHOOSE THE NEW **FERGUSON TRACTOR!**

What *is* the job-flexibility of a tractor for your farm? It's the total working range—performance *there for you to use* in all the many different power requirements, with different implements to handle—each job to be done with the very minimum of cost, the greatest possible saving of labor.

A large order only the New Ferguson Tractor can fill. Job-flexibility of the new Ferguson starts with the lighter jobs like planting, cultivating, haying and hauling, and belt work like shelling corn, grinding feed. Handy and cheap to run as a *small* tractor *which it is*. But just as capable—and a lot more economical—right at the top *and above* its working class—on heavy field jobs where full "2-plow-plus" power and pull must be produced.

Your needs on the farm—and Ferguson's years of experience in farm machinery—have developed and perfected the New Ferguson Tractor to its present, broad, all-inclusive job-flexibility, by means of the one and only Ferguson System of implement attachment and hydraulic control.

Measure the crop-to-crop and job-to-job performance of the New Ferguson against any other tractor. You'll choose the New Ferguson with Ferguson System advantages of job-flexibility and low operating costs. See your friendly Ferguson dealer for a demonstration.

Powerful Valve-in-Head CONTINENTAL ENGINE FOR JOB-FLEXIBILITY

Four speeds forward, each with a wide range of working power. High torque design provides steady flow of power even when running slow in high or third. "Wet" sleeves fully water-jacketed, improved cooling fan, full pressure lubrication, built-in oil filter, fully enclosed governor—precision manufacture throughout, for long life and low-cost operation in hard use.

FOR A BETTER LIVING AND A BETTER WORLD THROUGH . . .
Lower production costs and increased profits for the farmer . . . Lower food costs for the consumer . . . Less world unrest from hunger and want . . . Greater security for world peace.

FERGUSON TRACTOR AND FERGUSON SYSTEM IMPLEMENTS

For the one and only Ferguson System, look for this nameplate.

Harry Ferguson, Inc., 3639 E. Milwaukee Ave., Detroit 11, Mich.

Copyright 1949, by Harry Ferguson, Inc.

Ferguson

Ploughing with a narrow track setting, or inter row crop cultivation with a wide setting, highlights job flexibility of the Ferguson tractor and Ferguson System. April 1949.

COUNTRY GENTLEMAN

Model TO-20 Ferguson Tractor Illustrated

Measure PERFORMANCE...

YOU'LL CHOOSE THE NEW **FERGUSON TRACTOR!**

If you measure tractor performance by power, you'll get a big surprise when you feel the amazing responsiveness of the New Ferguson Tractor with its powerful, thrifty *overhead valve* engine.

But that's only part of the story! In the Ferguson, it's the revolutionary Ferguson System —*plus* the engine that gives this modern farming machine its great work-capacity . . . its ability to cut farm production costs so greatly.

The Ferguson System is built right into the Ferguson Tractor. It makes use of natural forces set in motion the moment a Ferguson Plow or other *unit* implement is pulled into and through the soil. And, best of all, these forces *automatically* adjust and adapt the Ferguson Tractor for heavy or light work, increasing traction as required.

And the Ferguson System of Hydraulics and 3-Point Implement Attachment lets you spend *more* of your time getting work done . . . *less* in bothersome manual control of implements . . . less in attaching and detaching of ground working tools. With *both* Finger Tip and Automatic Hydraulic Control, Ferguson Implement operation is truly effortless.

Before you buy a new tractor, measure the crop-to-crop and job-to-job performance of the New Ferguson against any other machine. You, too, will choose the New Ferguson Tractor with Ferguson System. Make a date now with your friendly Ferguson Dealer for *your* demonstration!

More Power

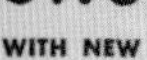

WITH NEW

Valve-in-head Engine

Open the throttle under load—you'll literally feel the extra power of this remarkable engine—jointly designed by Ferguson and Continental engineers. Exceptionally high torque at low engine speed gives amazing lugging power—often lets you use third speed when you'd expect to need second. Rugged, heavy-duty construction with drop-forged crankshaft, long-skirted, cam-ground aluminum alloy pistons, "wet" sleeves fully water-jacketed, full pressure lubrication and precision manufacturing—it's an engine built to "take it".

FOR A BETTER LIVING AND A BETTER WORLD THROUGH . . .

- Lower production costs for the farmer
- Lower food costs for the consumer
- Less world unrest from hunger and want
- Greater security for world peace

FERGUSON TRACTOR AND FERGUSON SYSTEM IMPLEMENTS

Harry Ferguson, Inc., 3639 E. Milwaukee Ave., Detroit 11, Michigan

Copyright 1949, by Harry Ferguson, Inc.

To make sure the tractor you buy has the one and only Ferguson System, look for this nameplate.
Ferguson

Why is that man not measuring the furrow depth with his Ferguson wrench? It was deliberately marked for jobs like this. Note that the rim to dish fixing brackets on the TO20 tractors were different from the TE20 tractors. February 1949.

COUNTRY GENTLEMAN

A DANGER POINT – WHEN IMPLEMENT MEETS OBSTRUCTION

Here's STOP-ACTION Protection that only Ferguson gives you!

Here is timesaving and moneysaving protection you won't want to be without in the next tractor you buy.

Farm losses from implement breakage are untold. You can take your own experience as a fair sample. But—safety and protection of implements can be taken for granted with the New Ferguson Tractor—just as you can take for granted the performance of its powerful valve-in-head engine built by Continental.

Whenever the implement strikes a rock or root, the Automatic Overload Release is touched off instantly and the Ferguson Tractor stops—always ahead of harm—never any damage done. *No other tractor has this feature!*

Another thing about the exclusive Ferguson "Stop-Action" protection you'll like is the ease and speed with which you can continue working the field. All you need do is back and raise the implement by Finger Tip Control, clear the obstruction, and go ahead. No unhitching and hitching—no lugging and tugging—no circling around to get lined up—no time lost.

But this is just one of the many things you'll like about the Ferguson System. Others are quick implement change and Finger Tip Control. Yes, and you'll like the way it maintains *greater* lugging power—not just at high or low—but over a *broad* range of speeds. All can be *shown* better than explained.

So, before you invest your money for ten, twenty years or more, be sure to see why the Ferguson Tractor pays for itself in savings.

Your nearest Ferguson dealer will be glad to demonstrate this modern tractor on your farm. Get in touch with him. He is a good man to know.

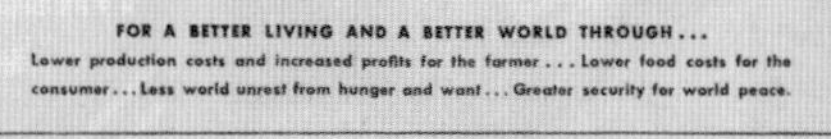

Ferguson

FERGUSON TRACTOR AND FERGUSON SYSTEM IMPLEMENTS

HARRY FERGUSON, INC., 3639 EAST MILWAUKEE AVENUE, DETROIT 11, MICHIGAN

Rare to see a Ferguson tractor skidding but it is designed to do this in these circumstances. This built in facility for wheel slip saved a great number of lives by stopping the tractor overturning backwards when the implement hits an obstruction. August 1949.

57
Ferguson Presents..
Another Engineering Achievement for Agricultural Progress
THE NEW FERGUSON MANURE SPREADER and FERGUSON LOADER
The Only One-man, One-tractor Manure Handling Combination!
Again the Ferguson System banishes a backbreaking farm chore!
Now you can handle TONS of manure easily, quickly with this combination . . . without ever leaving the tractor seat!
Load full, power-driven forkfuls by Finger Tip Control. Hitch and unhitch the Ferguson Spreader, too, with the same effortless Finger Tip Control. No more muscle-straining, time-consuming loading . . . no fussing with bothersome hand-hitches. It's all done by one man, one tractor and the Ferguson Manure Spreader and Loader.
BUILT TO SERVE AND SAVE
Built to Ferguson's high standards of quality, typical features include:
Spreader . . . Full 70-bu. capacity, single lever control, roller chain drive, double flight helicoid distributor, low loading height, corrosion-resistant covering.
Loader . . . Fast, easy one-man attachment, 1000-pound initial lift, foot trip release, tension-spring fork return, operates from the tractor hydraulic system, low over-all height for easy clearance.
Special hitch available for standard drawbar tractors.
Spreader is hitched and unhitched by a patented hook'n'eye operated by the Finger Tip Control Lever.
Six to ten power-driven forkfuls will load 70 bushels into the Spreader . . . all accomplished by Finger Tip Control.
Single control synchronizes beaters and conveyors to give accurate control of application per acre.
ASK YOUR FERGUSON DEALER FOR A DEMONSTRATION
WRITE FOR THE FULL STORY TODAY!
For the complete story of the timesaving, laborsaving Ferguson Manure Spreader and Loader, write Harry Ferguson, Inc. for your free copy of the 12-page, fact-filled booklet. (Address below.)
FERGUSON TRACTOR
Ferguson
AND FERGUSON SYSTEM IMPLEMENTS
COPYRIGHT, 1949, BY HARRY FERGUSON, INC., 3639 EAST MILWAUKEE AVENUE, DETROIT 11, MICHIGAN

The first Ferguson muck loader was introduced to the USA farmers before those in the UK. Note the different style of the USA pick up hitch compared to the UK version. November 1949.

The first Ferguson disc harrows were trailed, but the angle of the disc gangs was adjusted by means of a special hitch to the tractor's lower hydraulic linkage arms. May 1949.

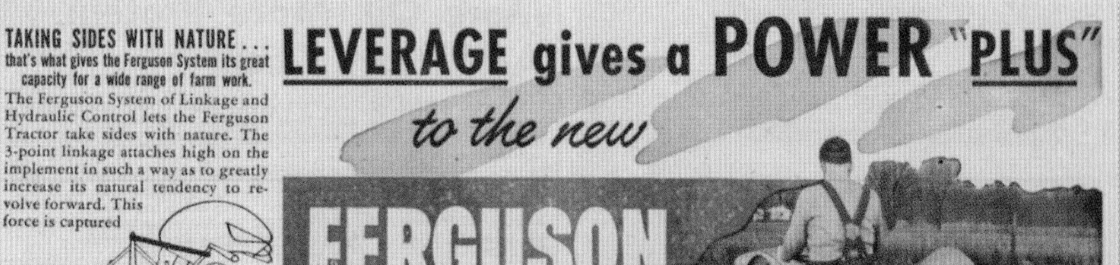

Ferguson at one with the forces of nature. I personally think that this description of the action of the Ferguson System is one of the best layman's versions that there is. 1949.

A revolution at 10 mph. Move your hay sideways and not forwards. Don't punish the leaves. April 1950.

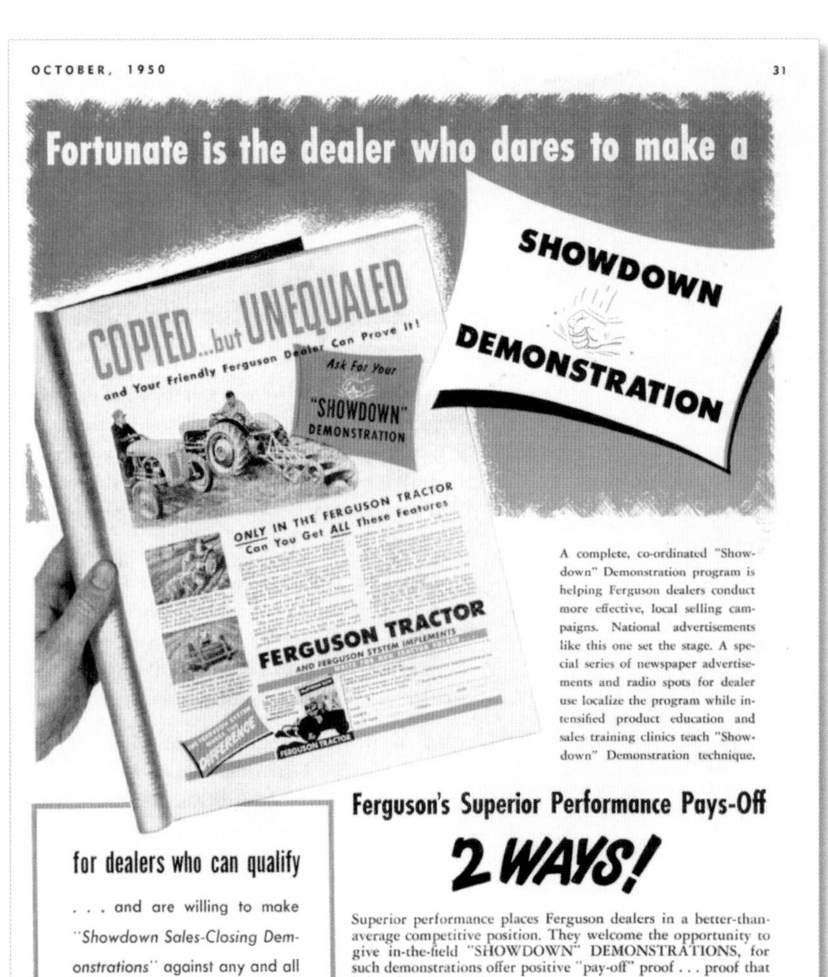

Ferguson was willing to take on any competition at "Showdown" demonstrations. It is reported that some small manufacturers would go into such demonstrations, knowing they could not compete with Ferguson equipment performance, just to get the wide publicity that these Ferguson events attracted. October 1950.

The driver has his gloves on. Fergusons never were a warm tractor to drive like some of the other more closed in ones of earlier times. Strange that in his quest for output per man Harry Ferguson never thought of making the driver more comfortable by offering a good Ferguson cab. Note how here we see Ferguson side badges on the hood for the first time. January 1951.

Yes Mr Ferguson. The whole world continues to benefit from your willingness to doggedly pursue a revolutionary design concept for implements and tractors.

"Most Copied". Is this a reference to Ford copying Harry Ferguson's designs after they terminated their business relationship?

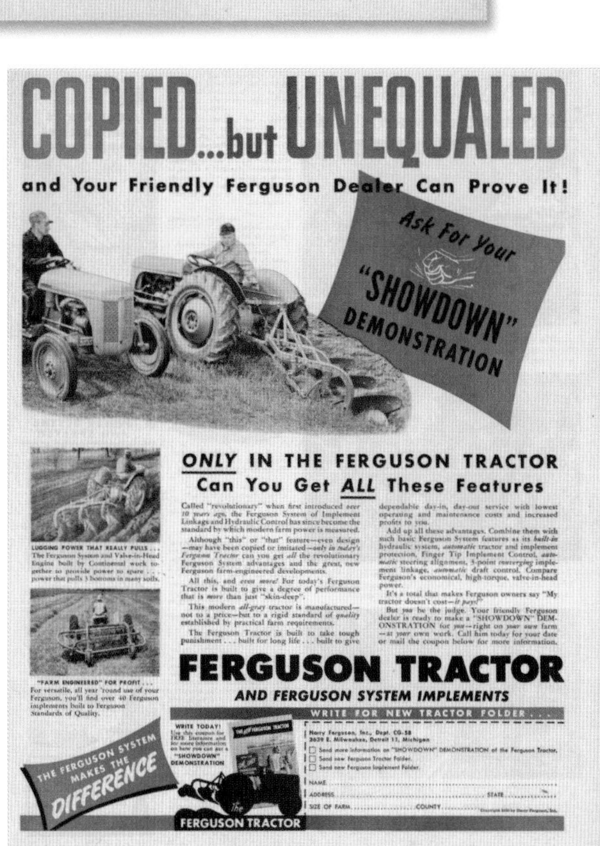

A challenge to Ford? Come and compete with a pure Ferguson at a Showdown demonstration.

Stressing the difference between Ferguson and the opposition. Note how the USA tractor had foot plates as standard whereas they were always optional in the UK. 1951.

COUNTRY GENTLEMAN
PLOWING
like this... takes
MORE than Power!
Get this...
The Ferguson Tractor uses one third to one half less power than ordinary tractors to do heavy jobs like this. Here's proof that the Ferguson has something more than brute power.
The big difference is the Ferguson System!
Go watch a Ferguson at work. See how this System transfers weight to provide extra traction when needed . . . how it gives heavyweight performance for heavy jobs, lightweight economy for others.
See how it keeps the front end down; how the built-in hydraulic mechanism controls working depth. See, too, how easily the finger-tip control can change working depth, or lift the plows at the headland furrow . . . how tightly the tractor makes its headland turn.
Listen to the snarl of the mighty engine—a specially designed valve-in-head engine that squeezes every bit of power from fuel.
STEP UP FOR YOUR "SHOWDOWN"!
Ask your Ferguson Dealer to stage a "Showdown" Demonstration right on your farm. He'll gladly match the Ferguson against any other tractor you may be considering. He'll prove to you . . . on your land . . . on your jobs . . . against any competition you select . . . that the Ferguson is the best tractor your money can buy! Harry Ferguson, Inc., 3639 E. Milwaukee Ave., Detroit 11, Mich.
Copyright 1951 by Harry Ferguson, Inc.
FERGUSON TRACTOR
AND FERGUSON Ferguson SYSTEM IMPLEMENTS

Worm's eye view of the plough in progress. What a superb claim that the Ferguson can use 33-50 percent less power.

COUNTRY GENTLEMAN
63 IMPLEMENTS
go on or off...fast!
Most all of the 63 Ferguson Implements go on or off in a jiffy . . . to increase tractor use . . . to save hours when switching from job to job.
It is these implements, just as much as the tractor, that determine how much work you can squeeze into busy days . . . how many different jobs you can do between morning and evening chores . . . how much more profitably you can farm your land.
The important thing to remember about these pace-setting Ferguson Implements is that they are designed for use with either today's advanced design Ferguson Tractor or with earlier tractors incorporating the Ferguson System. No Ferguson Implement has ever become an "orphan" because of design changes in a Ferguson Tractor!
Ferguson System Implements are built stronger to make full use of the tremendous power of the Ferguson's mighty valve-in-head engine. And they are constructed lighter to cut fuel and operating costs. All of them give you full benefit of the one and only Ferguson System.
Step Up for Your "Showdown"
It will pay you to examine some of these implements at your Ferguson Dealer's. Go over them point by point. Arrange for a demonstration on your farm. If you have doubts about the superiority of the Ferguson Tractor and Ferguson System Implements, arrange a "Showdown" demonstration so that you can see competing implements matched against each other on the same jobs. Your Ferguson Dealer will gladly co-operate. Harry Ferguson, Inc., 3639 East Milwaukee Avenue, Detroit 11, Michigan.
Copyright 1951 by Harry Ferguson, Inc.
Ferguson
FERGUSON TRACTOR
AND FERGUSON SYSTEM IMPLEMENTS

By the end of the production run of the TO20, we see an amazing 60+ implements being offered by Ferguson for the tractor. These were handed on to its successor, the TO30. July 1951.

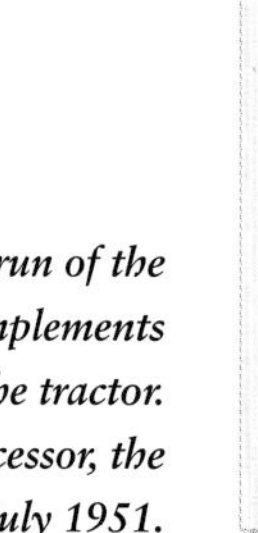

THE FERGUSON TO30

The evolution of the TO20 to the TO30 in effect mimicked the change from an 80mm engine to an 85mm engine in the TE20 range. However in the TE20 range there was no such re-designation. The TO30 had a larger Continental Motors engine than the TO20, hence a little more power and weight. Otherwise it was essentially unchanged. A strange feature of the advertising of the TO30s was that considerable emphasis was laid on the tractor having "suction side control" to the hydraulic system. The reason for the emphasis on advertising this feature are far from certain since it had been applied to Ferguson tractors since the Ferguson Brown. It is also significant that this feature was never noted in any UK advertising – it was purely a feature of the USA advertising of this particular model. It is postulated that maybe Ferguson felt the need to press home the advantages of the design of his hydraulic system compared to that of other tractor manufacturers. Ferguson was using a four-piston pump with suction side control whereas other manufacturers were almost without exception using gear pumps and control on the output side. The Ferguson System pump arrangement gave a far smoother and gentler operation.

Here the new TO30 has gained the "30" designation on the Ferguson badge on the side of the hood. The big front tyres make it look more bold. September 1951.

Beneath the sleek hood of the Ferguson "30" is a mighty valve-in-head engine designed to produce over twenty per cent more power than any previous Ferguson engine. It does far more work on each gallon of fuel. It runs far longer and far more efficiently between overhauls.
No other tractor ever produced has a better *torque** characteristic (what you call *lugging power*). Where other tractors falter or stall, or are forced to a lower gear, the Ferguson "30" pulls right on through . . . to give you greater operating convenience, faster work, savings in fuel and engine wear.

Give the old TO20 a 20 per cent bigger engine and wider front tyres and you have the new Ferguson 30. 1951.

HERE IT IS . . .
with even greater
TORQUE*
. . . the great new FERGUSON "30"
All over America, farmers are witnessing the most amazing performance in tractor history. Much of the greatness of the new Ferguson "30" is due to the revolutionary Ferguson System. Some of it to a 20 per cent increase in the power of its great new valve-in-head engine.
But the really outstanding feature of this great new tractor is its *torque* performance . . . what *you* usually refer to as "lugging power". Torque is vitally important to the man who's buying a tractor for his farm power requirements.
With the new Ferguson "30", you don't have to worry when the going gets heavier and heavier. The big reserve of lugging power pulls you through with the greatest of ease . . . with no shifting of gears . . . no stopping to "rev" up the engine . . . no danger of stalling. You'll keep right on going where other tractors would quit!
Many, many advanced engineering features have been built into the new Ferguson "30" to give you greater value for your money. Such as rotating valves that can't stick or burn. New, big, rugged gearing that provides far greater durability and longer rear-end life. A new air-fuel system that gives more work from each gallon of gas.
ASK YOUR DEALER FOR A DEMONSTRATION
A telephone call to your dealer will set up a demonstration right on your own land . . . with *you* at the wheel of a Ferguson "30". Once you've done a job with the Ferguson "30", you'll never let your dealer take it away. Call him *now!*
Building a better world through better farm mechanization is the *exclusive* business of Harry Ferguson, Inc., Detroit 32, Michigan.
*TORQUE EXPLAINED
Write for your FREE copy of "The Axe and the Wrench", the simple story of Torque and what it means to you in daily work with your farm tractor.
FERGUSON TRACTOR
AND 63 FERGUSON SYSTEM IMPLEMENTS

Not just more power for the TO30, but some extra torque too. Looks like a great day for the ploughman. Has anyone out there found a copy of that booklet on torque? No doubt Ferguson would have managed to get the subject into simple language. 1951.

Can you resist the temptation of a test drive and a Showdown demonstration? 1952.

Three 14 inch furrows is no mean feat for a tractor of this size. But we have three happy converts to the Ferguson System. April 1952.

Now...only the
FAR MORE POWERFUL
FERGUSON "30"
can have the famous
FERGUSON SYSTEM
with
Exclusive SSC*
SUCTION-SIDE CONTROL
BIGGER
★ in power
★ in performance
★ in economy
* As a result of the consent judgment in the recent settlement of the Ferguson vs. Ford lawsuit, Ferguson, and only Ferguson, can now have the famous Ferguson System. If you want the advantages of the Ferguson System, the Ferguson "30" is the only tractor that can give them to you.
Remember this . . . the Ferguson System is far more than 3-point linkage. True, this linkage is a part of the Ferguson System that can be seen. But, the Ferguson System control center is inside the tractor, the very heart of the machine.
There are other tractors on the market that are and will be similar in outward appearance to the Ferguson. Don't be fooled by these outward appearances! A vital and exclusive feature of the Ferguson System—the hydraulic pump with Suction Side Control—is buried deep within the tractor.
Your Ferguson Dealer will arrange a "Showdown" demonstration of the new and far more powerful Ferguson "30" on your farm. He will gladly prove to you the advantages of the Ferguson System with exclusive Suction Side Control.
The one and only, complete
FERGUSON SYSTEM
gives you these 5 important features:
1. Penetration without excess built-in weight
2. Traction without excess built-in weight
3. Finger tip and automatic draft control
4. Tractor's front end stays down
5. Automatic protection against hidden underground obstructions
This Booklet Tells the Story
Important facts about the Ferguson System and Suction Side Control are clearly and simply explained in this new booklet. Ask your Ferguson Dealer for a copy of "The Inside Story of the Ferguson System with Exclusive Suction Side Control".
Harry Ferguson, Inc., Detroit 32, Mich.
FERGUSON SYSTEM
SUCTION SIDE CONTROL
FERGUSON TRACTOR
and 63 Ferguson System Implements

Having won the lawsuit with Ford over their infringement of patents, Ferguson is now keen to capitalise on the regained sole ownership of the Ferguson System. August 1952.

The significance of Ferguson "Suction Side Control" was that by controlling low pressure oil flow into the Ferguson piston-type hydraulic pump (rather than controlling high pressure flow on the outlet of a conventional gear pump) several important benefits were gained. These were a) smooth, more controllable and gentle lift; b) no engine power used when not lifting; c) avoids excessive heating of oil; d) longer pump life; e) piston-type pump can have lower engineering tolerances and hence less critical oil filtration requirements. October 1952.

Same message, a different driver. 1952.

Now it's back to emphasising the power of that new engine. 1952.

At last the new mounted tandem discs make an appearance. And Ferguson will even advise you on how to buy a tractor – how good of them! 1953.

This and the next three advertisements all use photos from the same shoot but supported by different messages. See and believe. 1953.

March 7, 1953
THE FARMER
43
There's only one way... SEEING IS BELIEVING!
Get a SHOWDOWN DEMONSTRATION of the Far More Powerful FERGUSON "30"
BIGGER
★ in power
★ in performance
★ in economy
Yes . . . there's only one way to buy a tractor. And that's to see it demonstrated on your own farm doing *your* jobs under *your* conditions!
It doesn't matter what type or size tractor you own. It doesn't matter what other makes of tractors you've looked at. You'll get a completely different impression of power, performance, economy . . . of downright *value* . . . when you see a Ferguson at a Showdown Demonstration.
No descriptive folders or advertisements, in fact nothing a Ferguson Dealer can tell you, can do full justice to the facts about Ferguson performance . . . and what this tractor can do to cut your operating costs. Only an actual on-the-farm demonstration can do this.
Once you've had a Showdown Demonstration . . . once you've seen how well a Ferguson does *many* different jobs on *your* farm . . . you can judge this performance by any yardstick you choose. This may be your present tractor, or any make tractor.
But, regardless of the comparison you make, you'll be certain that Showdown Proof helps you buy more wisely. Why not ask your Ferguson Dealer for a Showdown Demonstration real soon?
Ask your Ferguson Dealer for Showdown Proof of unmatched performance on . . .
MOLDBOARD PLOWING
DISC PLOWING
FIELD TILLING
SUBSOILING
ROTARY HOEING
WOOD-LOT SAWING
MIDDLEBUSTING
MOWING
HARROWING
LOADING—SPREADING
ROW-CROP CULTIVATION
SIDE-DELIVERY RAKING
PLANTING
DITCHING
TERRACING
HAULING
Get your free copy of "How to Buy a Tractor" from your friendly Ferguson Dealer.
TRACTOR
DISTRIBUTED BY
MOTOR POWER EQUIPMENT CO.
St. Paul, Minnesota
2446 University Avenue
Phone NEstor 4544

Get that tractor demonstrated under your conditions. March 1953.

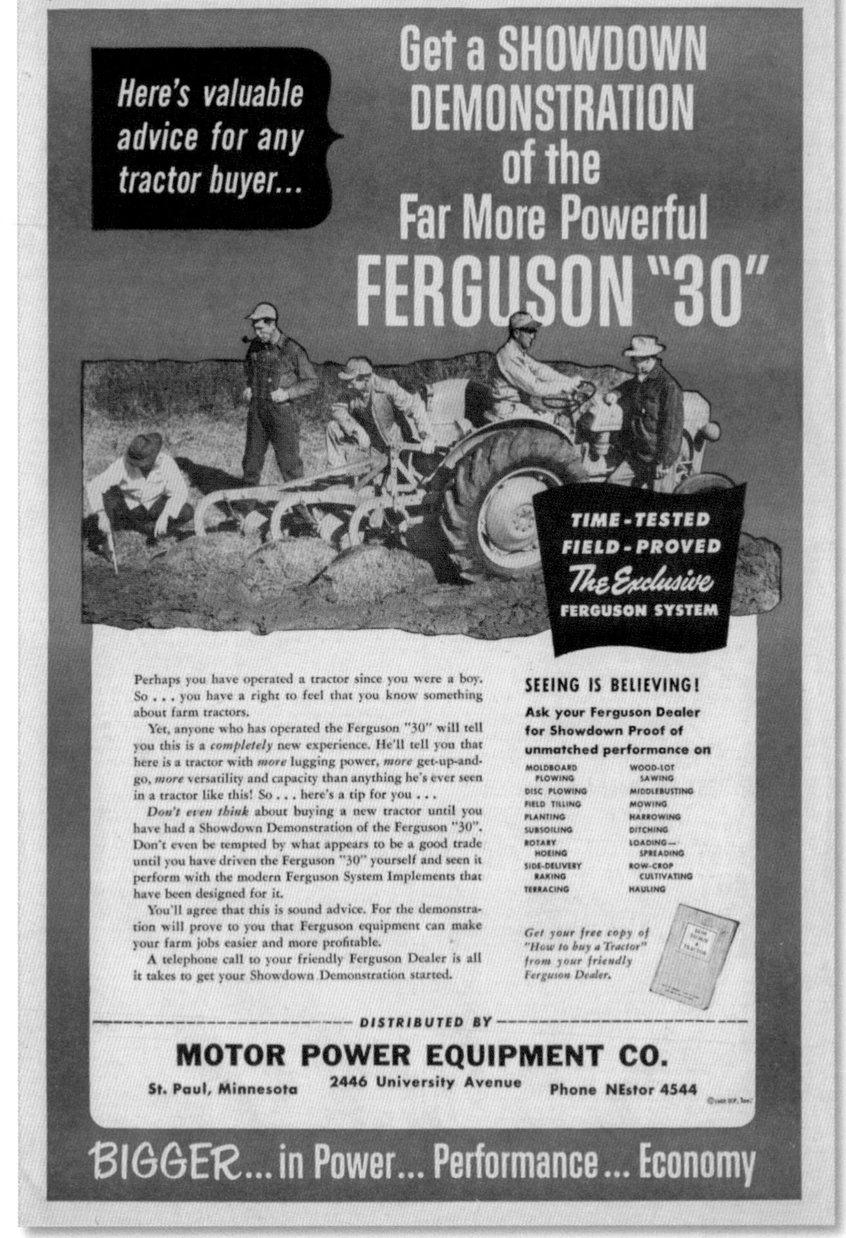

April 4, 1953
THE FARMER
33
Here's valuable advice for any tractor buyer...
Get a SHOWDOWN DEMONSTRATION of the Far More Powerful FERGUSON "30"
TIME-TESTED
FIELD-PROVED
The Exclusive
FERGUSON SYSTEM
Perhaps you have operated a tractor since you were a boy. So . . . you have a right to feel that you know something about farm tractors.
Yet, anyone who has operated the Ferguson "30" will tell you this is a *completely* new experience. He'll tell you that here is a tractor with *more* lugging power, *more* get-up-and-go, *more* versatility and capacity than anything he's ever seen in a tractor like this! So . . . here's a tip for you . . .
Don't even think about buying a new tractor until you have had a Showdown Demonstration of the Ferguson "30". Don't even be tempted by what appears to be a good trade until you have driven the Ferguson "30" yourself and seen it perform with the modern Ferguson System Implements that have been designed for it.
You'll agree that this is sound advice. For the demonstration will prove to you that Ferguson equipment can make your farm jobs easier and more profitable.
A telephone call to your friendly Ferguson Dealer is all it takes to get your Showdown Demonstration started.
SEEING IS BELIEVING!
Ask your Ferguson Dealer for Showdown Proof of unmatched performance on
MOLDBOARD PLOWING
DISC PLOWING
FIELD TILLING
PLANTING
SUBSOILING
ROTARY HOEING
SIDE-DELIVERY RAKING
TERRACING
WOOD-LOT SAWING
MIDDLEBUSTING
MOWING
HARROWING
DITCHING
LOADING—SPREADING
ROW-CROP CULTIVATING
HAULING
Get your free copy of "How to buy a Tractor" from your friendly Ferguson Dealer.
DISTRIBUTED BY
MOTOR POWER EQUIPMENT CO.
St. Paul, Minnesota
2446 University Avenue
Phone NEstor 4544
BIGGER... in Power... Performance... Economy

Listen to the experiences of those who already own a Ferguson. April 1953.

The Ferguson dealer will help you make an important decision. 1953.

It makes sense to choose your tractor this way

Base your decision on *performance*—not talk.

To help you do this, Ferguson has devised the conclusive Showdown Demonstration. It is a *performance* test right on your own farm. It proves just how well the powerful Ferguson "30" will do most of your jobs—*more* of the time than any other tractor!

At your own Showdown Demonstration, you'll see the rugged Ferguson perform your toughest jobs without effort, your easier jobs without wasting power. You'll be able to compare the performance with that of your present equipment, or that of any other make you're considering.

So take this sensible step now . . . Call your Ferguson Dealer and arrange for your Showdown Demonstration of the powerful Ferguson "30". There's no obligation. You'll make your own decision about the best tractor for *you*—based on performance alone, not claims!

What you need to know about tractor buying! Ask your Ferguson Dealer for free copy of 24-page, fact-packed booklet, *How to Buy A Tractor*, or write: Harry Ferguson, Inc., Detroit 32, Michigan.

A Showdown Demonstration will show you remarkable new Ferguson developments, like the Tiller, for low-cost seedbed preparation. You'll also see how *convenient* the Ferguson is, with *fast* hook-up, *eyes-ahead* steering, and tractor-mounted implements!

No other tractor gives you the Ferguson System! That means traction without power-stealing weight, plus front-end stability, and a simple built-in hydraulic release that saves your tractor and implement if a hidden rock or stump is hit. Get *all* the facts about the Ferguson System at *your* Showdown!

Seeing is Believing — Get Your Showdown Demonstration of the

FERGUSON "30"

A good shot of the disc plough at work in conditions where it is favoured – hard and dry. One of the last advertisements before the merger with Massey-Harris. Autumn 1953.

Changing times. Note how the address for Ferguson has moved to Racine, Wisconsin. The merger with Massey-Harris has occurred now. 1954.

In most soils, probably *yours,* the Ferguson "30" will turn three furrow slices easily and smoothly! We'll prove to you . . .

Why Ferguson can be a 3-Plow Tractor

• It's because of the Ferguson System. It makes a single unit of the plow and tractor by means of converging 3-point linkage and a built-in hydraulic controlling mechanism.

As you plow with this integrated unit, the Ferguson System actually makes use of the plow's weight, plus the weight of the soil forces on the plow. It automatically *adds* tractive weight to the tractor, as you need it—through the *hydraulic system.* On light jobs, when you don't need as much tractive weight, it's not there to waste your fuel!

That makes Ferguson different! Even though other tractors may hook up at 3 points and have some form of hydraulic system.

We can understand it if you want proof that the Ferguson has the power and weight to do your big jobs. You deserve that proof . . . and you can get it by taking advantage of the offer below.

Yours for the asking—a sample of Ferguson System farming! Print your name and address in the margin and mail this advertisement to: *Ferguson,* Racine, Wis. We'll send an expert to your farm to give you a sample of the Ferguson "30" doing *your* big jobs. Write now!

YOU'LL SEE MORE AND MORE OF THE

Ferguson "30"

SEE YOUR FERGUSON DEALER FIRST!

16

Is the reference to the big jobs now a challenge to Ferguson's new partner Massey-Harris in this period of a dual line marketing policy? March 1954.

22 THE FARMER March 6, 1954

Get accurate planting at high speed with the tractor-mounted Ferguson Drill Planter! Giant-sized, ½-bushel seed hoppers mounted low for easy filling and shorter seed drop. With finger tip control you raise and lower the planter hydraulically for easy transport and clean head-land turns. Ask your Ferguson Dealer!

Attached in 60 seconds, the Ferguson Rigid Tine Cultivator is rear mounted and finger tip controlled for operating ease and accurate depth control. Heavy-duty tines work together as a unit for thorough weed killing and soil mixing action. And the Ferguson System linkage plus front-end guide gives you accurate *eyes ahead* cultivation. Ask your Ferguson Dealer!

How Ferguson gains weight for big jobs

Ferguson System Implements actually become part of the Ferguson Tractor . . . through a simple, 3-point linkage and unique, built-in hydraulic controls.

As a result, resistance of the soil against ground-engaging tools is put to work for you, adding weight to the tractor, as needed, to produce more traction.

When soil resistance increases in heavier soils, the tractor automatically gains hundreds of pounds of extra traction weight through the Ferguson System. As resistance decreases, traction weight is lessened.

Right there's where Ferguson is different!

Different even from other 3-point hookups and hydraulic systems. It's how the compact Ferguson "30" can pull three bottoms with ease in most soils . . . probably *yours!*

But when the going is easy, as it is on *most* farm jobs, the Ferguson System doesn't waste fuel on excess built-in weight.

If it's hard for you to see how the low-cost Ferguson has the power and weight for *your* big jobs, we can understand why. All we ask is that you read the offer on the right, and take advantage of it *today.*

How to get a sample of Ferguson System farming: Print your name and address in the margin of this advertisement and mail it to: *Ferguson,* Racine, Wisconsin. We'll have an expert report to your farm to give you a generous sample of the Ferguson "30" doing your big jobs. *Write today!*

Ferguson is the only tractor in the world being made with all the advantages of the Ferguson System: penetration and traction without excess weight, finger tip and automatic draft control, front-end stability, and fully automatic protection if implement strikes hidden obstruction. And remember . . . only Ferguson Dealers can give you the genuine Ferguson System!

YOU'LL SEE MORE AND MORE OF THE

Ferguson "30"

CALL YOUR NEAREST FERGUSON DEALER

How Ferguson can lug 3 plows

It doesn't look it—but the low-cost Ferguson "30" turns three furrows easily in most soils! How?

The Ferguson System's exclusive linkage and built-in hydraulic control makes one unit of tractor and implement. Tillage tools are mounted on the tractor, adding *weight*, adding *traction*.

Still more traction is provided by the soil's weight and resistance on the implement, plus the suck of the tool. These forces become "weight" and are carried to the rear wheels by the Ferguson linkage and hydraulic system.

In heavier soil, tractive weight is automatically added through the Ferguson System, *as needed*, by the additional resistance and weight of heavier soil on the tool!

And since the Ferguson System is constantly "weighing" this resistance, tractive weight is automatically decreased on lighter work. Thus, Ferguson's tractive weight always matches the *need* for weight. *You save money*—because you never pay for wasteful excess power or needless built-in weight.

We owe you proof that the low-cost, compact Ferguson "30" can be a 3-plow tractor. To get that proof, simply take advantage of the free offer below.

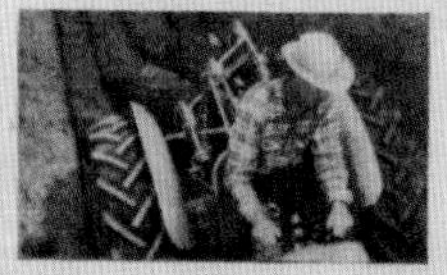

Get your sample of Ferguson System farming: Print your name and address in the margin and mail this ad to *Ferguson*, Racine, Wisconsin. We'll have an expert come to your farm and give you a generous sample of the Ferguson "30" doing *your* big jobs.

YOU'LL SEE MORE AND MORE OF THE

Ferguson "30"

SEE YOUR FERGUSON DEALER FIRST!

89

A demonstration on your farm is free. August 1954.

How Ferguson gains weight for big jobs

Ferguson System Implements actually become part of the Ferguson Tractor . . . through a simple, 3-point linkage and unique, built-in hydraulic controls. As a result, resistance of the soil against ground-engaging tools is put to work for you, adding weight to the tractor, as needed, to produce more traction.

When soil resistance increases in heavier soils, the tractor automatically gains hundreds of pounds of extra traction weight through the Ferguson System. As resistance decreases, traction weight is lessened.

Right there's where Ferguson is different! Different even from other 3-point hookups and hydraulic systems. It's how the compact Ferguson "30" can pull three bottoms with ease in most soils . . . probably *yours!*

But when the going is easy, as it is on *most* farm jobs, the Ferguson System doesn't waste fuel on excess built-in weight.

If it's hard for you to see how the low-cost Ferguson has the power and weight for *your* big jobs, we can understand why. All we ask is that you read the offer below, and take advantage of it *today*.

To get a sample of Ferguson System farming: print your name and address in the margin of this ad and mail it to: Ferguson, Racine, Wisconsin. We'll have an expert report to your farm to give you a generous sample of the Ferguson "30" doing your big jobs. *Write today!*

YOU'LL SEE MORE AND MORE OF THE

Ferguson "30"

SEE YOUR FERGUSON DEALER FIRST!

50

Bird's eye view of the Ferguson plough cutting a very fine, clean furrow. 1954.

How Ferguson adjusts weight to save on light jobs

The revolutionary Ferguson System transforms tractor and implement into a *single* power-farming *unit*.

It does this through the Ferguson-invented, three-point, sixty-second implement attachment, and a unique system of tractor-mounted implements with hydraulic draft control.

Which means that when you plow or do other *heavy* tillage work, the Ferguson System adds the soil's weight to the tractor-implement weight—through the hydraulic system. This automatically gives you *extra* traction weight when you need it . . . lets you pull three bottoms in most soils easily and smoothly!

But most important: On light work . . . the day-in, day-out jobs you do *most* of the time . . . your Ferguson isn't forced to haul excess, built-in weight that drinks your gas and oil. This is not true when you do light jobs with an *oversized* tractor.

Thus, Ferguson gives you big-tractor capacity, but does *not* give you big-tractor fuel bills on light work!

Are you interested in improving your income? Then we want to show you how the low-cost Ferguson "30" can do more of your jobs, more of the time, at less cost than other tractors. Read the offer below and take advantage of it today!

Get your sample of Ferguson System farming: Print your name and address in the margin of this ad and mail to: *Ferguson*, Racine, Wisconsin. We'll have an expert come to your farm and give you a generous sample of the Ferguson "30" doing your jobs at a saving. Write today!

YOU'LL SEE MORE AND MORE OF THE

Ferguson "30"

SEE YOUR FERGUSON DEALER FIRST

Weight transfer not required for all tasks.

TWO WORLD FIRSTS IN TRACTOR-MOUNTED IMPLEMENTS

Here are the most revolutionary advances in agricultural implements since the development of the self-propelled combine! Ferguson's new Forage Harvester and new Baler—tractor-mounted and semi-self propelled—operate with hitherto undreamed-of efficiency. Yet their cost is so low it assures you of a wide and ready market for sales.

In rigorous field tests the new Forage Harvester and new Baler (soon to be ready for distribution to dealers) proved successful beyond their designers' highest hopes. They have encouraged Ferguson engineers to work for still greater advances in other areas of farm implement design. That's why we can say with complete confidence . . .

You'll see More and More of Ferguson

The Ferguson Forage Harvester is mounted on the *side* of a Ferguson Tractor, by one man, in two to four minutes. The driver looks *ahead* to watch the Forage Harvester devour crops—standing or windrowed—at high speed. Its short turning radius and rigid mounting make it the most manoeuvrable of all forage machines.

The Ferguson Baler needs no heavy lifting or special tools for attachment, and the job is done in less than 90 seconds! And the farmer gets every advantage of a compact, highly manoeuvrable "self-propelled" baler. In turning radius, handling ease, rigidity of its coupling with the tractor, the new Ferguson baler is the equal of the amazing Forage Harvester.

SEE THIS REVOLUTIONARY HAYING EQUIPMENT AT THE INTERNATIONAL PLOWING MATCH, BRESLAU, ONT., OCTOBER, 12-15

MASSEY-HARRIS-FERGUSON LIMITED
TORONTO, CANADA

World firsts again for Ferguson advertised in Canada under the new Massey-Harris-Ferguson company name. September 1954.

SET YOUR STANDARDS HIGH

... Then ask for a Ferguson Demonstration

All it takes is a phone call! And at a time of your choosing, your Ferguson dealer will bring a lively new Ferguson tractor spinning up your driveway, ready for a try-out on your own farm.

Then *you* can sit behind the wheel, try it for power, comfortable handling, manoeuvrability. Drive it out on your own farmland and see for yourself why it can handle a three-bottom plow. There's a *real* test of the famous Ferguson System for you!

Just to give the dealer a hand, we'd like to explain why the Ferguson makes such light work of so many heavy jobs. The fact is, the Ferguson doesn't need built-in weight to do the work of much heavier tractors. It uses the pressure of the soil on the implement to establish greater traction, and greater front-end stability.

Tractor and implement virtually become one. Yet Ferguson's unrivalled hydraulic system automatically keeps the implement at just the right working depth. Strike a hidden obstruction, and the impact instantly releases the implement weight from the rear wheels. This allows them to spin harmlessly, prevents damage to implement and tractor, and avoids possible injury to the operator. Only Ferguson has this great safety feature.

The Ferguson is time-proven, farm proven. Better make that phone call soon and prove it for yourself.

LOWEST-PRICED TRACTOR OF ITS TYPE IN CANADA!

Today the Ferguson costs substantially *less* than any other tractor of its type! *No other* tractor gives you Ferguson's many advantages, and Ferguson's low price besides!

Massey-Harris-Ferguson
Toronto, Canada **LIMITED**

"This advertisement is one of a series currently appearing in Canadian Farm Papers."

CANADIAN FARM IMPLEMENTS—*April, 1955* 49

Up in Canada again this seems to be a TO30 in 1955 being advertised after production ceased in 1954. April 1955.

THE FERGUSON TO35

The American-built TO35 and UK-built FE35 were essentially the same basic tractors but the TO35 was released a year earlier, in January 1955, just five tractors having been produced in late 1954. Both had distinctly new liveries and tinwork design compared to the previous TE and TO tractors. Initially there were a few grey/green tractors followed by cream and grey tractors, and finally a red and grey version. Both had more powerful (larger) engines and were a little heavier than their predecessors. However the TO35 was produced mainly with a petrol engine until the British Standard 23C diesel engine was later offered as an option. Earlier tractors had 6V ignition but a change to 12V came with the later red and grey models.

There were major mechanical changes principally in the form of a dual range gearbox to give a total of six forward and two reverse gears, and a fundamental change to the hydraulic mechanism which became a two instead of one lever control. This facilitated draft and position control. A two stage clutch was also offered which facilitated "live PTO". Two speed pto was also offered – engine or ground speed. These were perhaps the main engineering design innovations but there were also many small changes that were common to the TO and FE35s. A significant one of these was to place the independent brake pedals on one side of the tractor instead of on separate sides as on the TE and TO tractors. These changes had been designed in Harry Ferguson's Detroit engineering design unit and not the UK. Like the previous TO20s and TO30s there were almost no model variants but a previously unknown military version has just been found.

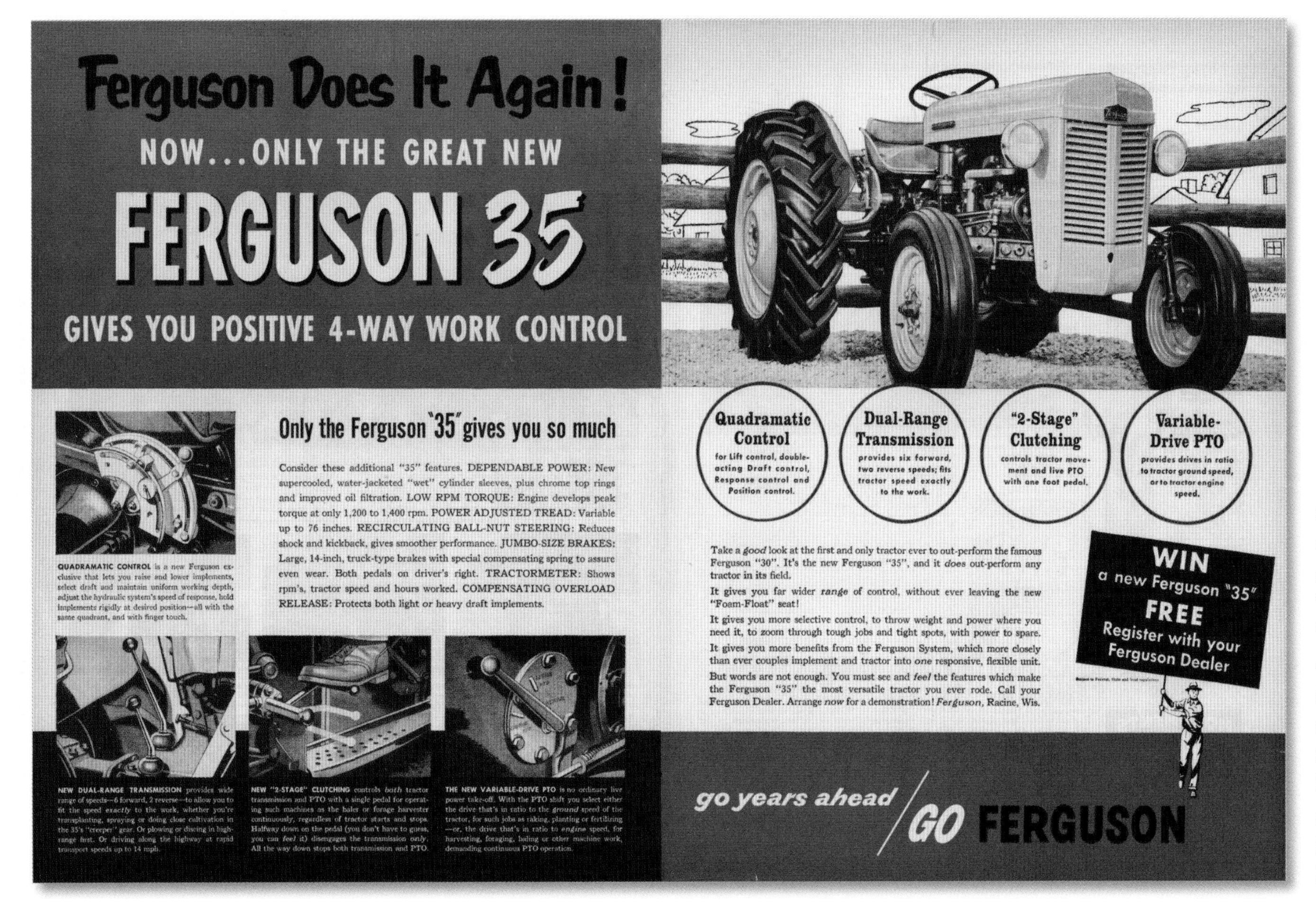

Only four TO35 tractors were made in 1954, so the advertising of them starts in earnest in 1955. The tractor was an improved TO30 having dual range transmission, quadramatic hydraulic controls, two stage clutch, and variable drive pto as the main design improvements. Note too how the old Ferguson chevron badge on the front of the hood has been given a new shape. February 1955.

make this test with your own tractor . . .

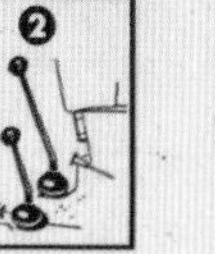

Here for the first time in *any* tractor is a combination of revolutionary engineering advances that will enable you to farm *more*, work *less*, realize greater profit from your work. But don't take our word for it. Look for *all four* of these work-control features on your present tractor. We believe you'll agree . . . only the Ferguson "35" gives you so much.

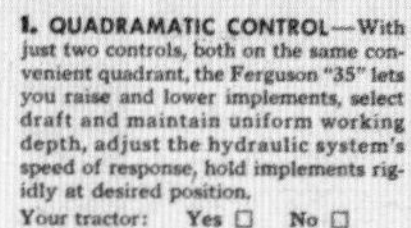

1. QUADRAMATIC CONTROL—With just two controls, both on the same convenient quadrant, the Ferguson "35" lets you raise and lower implements, select draft and maintain uniform working depth, adjust the hydraulic system's speed of response, hold implements rigidly at desired position.
Your tractor: Yes ☐ No ☐

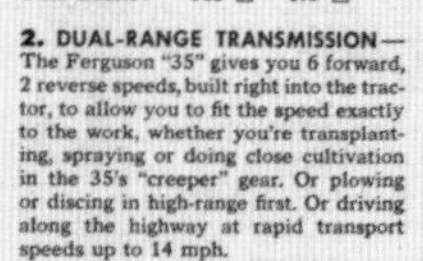

2. DUAL-RANGE TRANSMISSION—The Ferguson "35" gives you 6 forward, 2 reverse speeds, built right into the tractor, to allow you to fit the speed exactly to the work, whether you're transplanting, spraying or doing close cultivation in the 35's "creeper" gear. Or plowing or discing in high-range first. Or driving along the highway at rapid transport speeds up to 14 mph.
Your tractor: Yes ☐ No ☐

3. "2-STAGE" CLUTCHING—Means single-pedal control of both transmission and PTO for operating such machines as the baler or forage harvester continuously. Halfway down on the pedal (you don't have to guess, you can feel it) disengages the transmission *only*. All the way down stops both transmission and PTO.
Your tractor: Yes ☐ No ☐

4. VARIABLE-DRIVE PTO—The "35" offers much more than ordinary "live power take-off." With the PTO shift you select either the drive that's in ratio to the ground speed of the tractor, for such jobs as raking, planting or fertilizing—or, the drive that's in ratio to engine speed for harvesting, foraging, baling or other machine work demanding continuous PTO operation.
Your tractor: Yes ☐ No ☐

Call your local Ferguson dealer. Arrange now for a demonstration.
Ferguson, Racine, Wisconsin

WIN A NEW FERGUSON "35" FREE
REGISTER WITH YOUR FERGUSON DEALER

go years ahead / **GO FERGUSON**

MARCH 5, 1955 75

Two hydraulic levers, two gear levers, two clutch positions and two pto positions. The original simplicity of the Ferguson tractor for the last two decades lost forever. March 1955.

RIGHT NOW...

GET IT WITH THE FERGUSON 35

EXCLUSIVE 4-WAY WORK CONTROL
lets you farm more, work less

1. QUADRAMATIC CONTROL
for Lift control, double-acting Draft control, Response control and Position control.

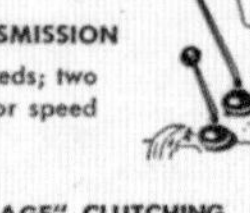

2. DUAL-RANGE TRANSMISSION
provides six forward speeds; two reverse speeds; fits tractor speed exactly to the work.

3. "2-STAGE" CLUTCHING
controls tractor movement and live PTO with a single foot pedal.

4. VARIABLE-DRIVE PTO
provides drives in ratio to tractor ground speed, or to tractor engine speed.

You've heard of the brilliant new Ferguson "35", of course. Now try it yourself . . . on your own farm. Get the feel, at first hand, of exclusive 4-Way Work Control at work *for you!*

And that's not the whole new story of this years-ahead tractor by any means. The "35" also gives you other advanced features such as High-Torque Engine . . . Recirculating Ball-Nut Steering . . . Sight-Glance Tractormeter . . . Compensating Overload Release. *Plus* many other Ferguson benefits.

Get set for a new appreciation of how flexible a tractor can be in all-around performance. How handy its fingertip operation. How selective and wide-ranging its controls. Seeing is believing . . . but a *tryout*, in person, is even more convincing. Call your Ferguson Dealer now to arrange just such a demonstration.

Ferguson, Racine, Wis.

Just a few more days to register with your Ferguson Dealer to **WIN A FREE "35"**
Register before April 1, 1955

go years ahead / **GO FERGUSON**

Many traditional Ferguson users did not agree that additional controls, in particular the two hydraulic levers, were a design advance and "years ahead".

More power, more control, more weight, more comfort

Here's the biggest news since modern farming took a giant forward step with the first Ferguson tractor!

Now Ferguson brings you a completely new and vastly improved tractor, the Ferguson 35. It's new under the hood, new in its method of implement control, new even in the smart, two-tone paint job. It's longer, heavier, and far more powerful.

The 35 brings you greater range, greater selectivity of implement control. Its engine has been redesigned to give high torque at low engine speed. This guarantees lugging power without surging or racing, less stalling from overloads, horsepower to spare for really tough jobs.

From the spacing of the rear wheels (48 to 76 inches) and the extra large, truck-type brakes to the new "Tractormeter" dial and "Foam Float" Seat, the new Ferguson 35 is truly an engineering triumph. You'll find it a wonder to drive . . . a real profit-builder to own.

New exclusive 4-way work control transforms power into profits!

1. **New Hydraulic System with Quadramatic Control** for Lift control, double-acting Draft control, Response control and Position control. Guarantees far greater accuracy for plowing, discing, cultivating.

2. **Dual-Range Transmission** provides six forward speeds, two reverse; fits tractor speed exactly to the work. Low range for harvesting, plowing, sub-soiling. High range (up to 14 mph) for light jobs or highway travel.

3. **2-Stage Clutching** controls tractor movement and live PTO with a single foot pedal. Pressing pedal half way down (you can feel when you're there) disengages transmission. All the way down stops both tractor and PTO.

4. **Variable Drive PTO** provides drives in ratio to tractor ground speed, or to tractor engine speed. Synchronize with engine for harvesting, with ground speed for jobs like raking.

Massey-Harris-Ferguson LIMITED

TORONTO, CANADA

FF-455T

"This advertisement is one of a series currently appearing in Canadian farm papers."

CANADIAN FARM IMPLEMENTS—*June, 1955* 25

Announcing the TO35 in Canada. This colour advertisement shows the tractor in its early grey/green livery. These "green bellies" are now highly sought after. June 1955.

Another PERFECT MATCH...

The New FERGUSON Dyna-Balance MOWER

No wonder farmers are choosing the new Ferguson Dyna-Balance Mower over the slow, noisy, outmoded pitman types. Exclusive Ferguson assembly balance ends vibration shakes . . . rugged new ground-level drive and special bearings increase knife speeds up to 30%. Knife-head wear plates are eliminated; lead adjustment and knife register are *factory set*. Virtually noiseless and vibration free . . . the Ferguson Dyna-Balance Mower means bigger production at lower cost, and far less wear and tear on the machine, on *you*.

Ferguson, Racine, Wisconsin.

Why is the new Ferguson "35" already such a solid No.1 favorite with farmers? For one reason, it's the *only* tractor combining such features as High Torque Engine, Double-Acting Control Spring, Sight-Glance Tractormeter and Compensating Overload Release (plus many other extras), with *exclusive 4-Way Work Control* . . . .

A new FERGUSON exclusive and the most talked-about TRACTOR advance in years

QUADRAMATIC CONTROL—For Lift control, double-acting Draft control, Response control and Position control.

NEW DUAL-RANGE TRANSMISSION—Provides six forward, two reverse speeds; fits tractor speed exactly to the work.

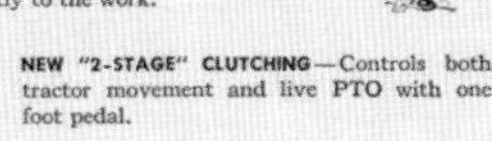

NEW "2-STAGE" CLUTCHING—Controls both tractor movement and live PTO with one foot pedal.

NEW VARIABLE-DRIVE PTO—Provides drives in ratio to tractor ground speed, or to tractor engine speed.

go years ahead / **GO FERGUSON**

CAPPER'S FARMER JULY 1955 33

A new tractor and a new mower. July 1955.

WHAT CONTROL MEANS
QUADRAMATIC CONTROL
VARIABLE-DRIVE PTO
DUAL-RANGE TRANSMISSION
"2-STAGE" CLUTCHING
IN THE NEW FERGUSON 35
with 4-way work control
Make no mistake, *control* is vital in modern-day mechanized farming.
Control means accurate, positive handling of all implements to assure relaxed, muscle-free operation right from the operator's seat. It means better seedbed preparation, more accurate planting and a properly timed harvest . . . at drastically *reduced* costs. It can mean premium crops and a greater profit. Simply: It puts farming on a paying basis—puts black ink in the account book.
And Ferguson leads the way in this business of *control:*
Its exclusive *Quadramatic Control* gives far more complete and accurate hydraulic control of implements . . . all with the same compact control quadrant. Its Dual-Range Transmission provides a wider range of speeds, geared exactly to the job demand.
A choice of two PTO drives is built in . . . one in ratio to ground speed (for raking, planting, fertilizing); one in ratio to engine speed (for forage harvesting, baling, mowing, etc.). And the operator controls *both* tractor transmission and PTO with a single clutch pedal.
The *combination* of these and other advantages, plus the famous Ferguson System of teaming tractor and implements, brings you a degree of control you'll find in no other tractor.
Ask your Ferguson Dealer for a demonstration . . . and learn why more and more farmers are being sold on the Ferguson "35" as *the Tractor* that offers the *most control* . . . by far, the most in all-out tractor value! *Ferguson*, Racine, Wisconsin.
BIG NEW CONTEST!
WIN A FERGUSON 35 FREE. You may win one of 26 FREE TRACTORS to be given away during July, August and September in the big "Better Days Ahead" Contest. SEE YOUR FERGUSON DEALER.
SUCCESSFUL FARMING, AUGUST, 1955

You can even win a Ferguson! This overhead shot shows the lines of the opening hatch in the hood. August 1955.

YOU FARM MORE—
WORK LESS
WITH THE NEW
FERGUSON
35

Uniform cultivation with Quadramatic Control

Smoother baling with live PTO

Faster raking with dual-range transmission

Easier materials handling with 2-stage clutching

Why? Because Only FERGUSON Gives you 4-Way WORK CONTROL

Power, lugging ability, precise implement control, low fuel bills, matchless handling ease—put them all together and you have the great *new* Ferguson 35! More than ever, with Ferguson's four new control features, you can make your implements perform *exactly* as you want them to. Here's why:

1. New Hydraulic System with Quadramatic Control—This Ferguson exclusive selects and maintains working draft automatically, lifts and lowers your plow without disturbing draft setting. The Double-Acting Control Spring measures both compression and tension forces, weight or draft, so you get greater accuracy for discing or cultivating. In utility jobs, such as operating a crane or loader, you can hold the implement rigidly in any position.

2. Dual-Range Transmission—New, low-range transmission gives full lugging power while maintaining low-range speed for tough harvesting, heavy-draft plowing or sub-soiling. Low range is useful, too, for close, accurate cultivation, transplanting or spraying. High range permits speeds up to 14 mph for mowing, raking, light hauling or highway travel. Six forward speeds, two reverse.

3. 2-Stage Clutching—Means single-pedal control of both transmission and PTO. For "live" PTO jobs, pressing the pedal half way down (you can feel when you're there) disengages transmission, but PTO continues to run for clearing such implements as the forage harvester and baler. Press the pedal all the way down and you can stop both tractor and PTO immediately.

4. Variable-Drive PTO—Gives drives in ratio to tractor *ground* speed, or to tractor *engine* speed. For harvesting, speed is synchronized with the engine; you harvest heavy stands quickly and economically. Put the PTO shift in "ground" position and you get a drive that's always in ratio to ground speed, ideal for operations like raking where uniform distribution of materials is required.

Massey-Harris-Ferguson
LIMITED
TORONTO, CANADA

"This advertisement is one of a series currently appearing in Canadian farm papers."

CANADIAN FARM IMPLEMENTS—*August, 1955* 41

Up in Canada again where the TO35 is demonstrating its versatility.

WEIGHT TRANSFER

Gives traction to work when you're ready ...with the all-season FERGUSON 35

4-Way Work Control Exclusive in the "35"

1. **Quadramatic Control:** Lift control, double-acting Draft control, Response control, Position control.
2. **Dual-Range Transmission** provides 6 forward, 2 reverse speeds; fits tractor speed exactly to work.
3. **"2-Stage" Clutching** controls both tractor movement and live PTO with one foot pedal.
4. **Variable-Drive PTO** provides drives in ratio to tractor ground speed or to tractor engine speed.

Plan now to level out your next year's work load. Forestall a *spring rush* with a *fall push!*

With the Ferguson Tractor and the famous Ferguson System—with its principle of weight transfer that provides needed traction *without excess built-in weight*—you can plan your fall work with confidence.

For example: If you're in the soft winter wheat area, you can get your corn off, prepare the land and seed winter wheat with a minimum of effort or delay. If you fertilize and plow in the fall, the all-weather Ferguson is ideal. If you use fall plowing strictly as a moisture trap, again the Ferguson lets you get into the field—to get work done on time.

And for off-season work, the Ferguson "35" offers endless benefits: terracing, ditching, leveling—with the Multi-Purpose Blade; loading, hauling, spreading—with the Manure Spreader and Loader; shattering hardpan, or laying plastic pipe or cable with the Subsoiler and Pipelayer.

So don't delay! Now is the time to swing into the seat of a new Ferguson "35" . . . to get the *feel* of 4-Way Work Control in operation. See your Ferguson Dealer, today.

Ferguson, Racine, Wisconsin.

26 FREE TRACTORS
CONTEST CLOSES SEPT. 30, 1955

Get in on BIG Contest. You may win one of these new Ferguson "35" Tractors FREE. See your Ferguson Dealer.

One of the first advertisements with the TO35 shown using PAVT wheels which enabled rapid changes of rear track width. 1955.

Another Years-Ahead
FIRST BY FERGUSON
4-Way Work Control on the FERGUSON 35
LIFT
POSITION
DOWN
FAST
RESPONSE
SLOW

How Ferguson Quadramatic Control Lets You Take Full Charge of Implements

Control of implements is sure and easy with a Ferguson "35" because of the time-tried and continually improved Ferguson System.

Right at your finger tips is new Quadramatic Control that will amaze you with its close command of implements. This Ferguson "first" lets you raise and lower implements, select draft and maintain working depth, adjust the hydraulic system's speed of response and hold implements at any position you choose—all with the same compact control quadrant.

And you'll find the other members of the Ferguson 4-Way Work Control team contributing their part to your mastery over every farm job: Variable-Drive PTO for versatile, synchronized power; "2-Stage" Clutching for one-pedal control of both transmission and PTO; and Dual-Range Transmission to insure peak efficiency in each speed range.

No longer need your tractor's limitations dictate how well you farm. Ask your Ferguson Dealer to show you how the Ferguson "35" lets you take *complete* charge!

Ferguson
Racine, Wisconsin

FOR PLOWING, draft is maintained automatically, even with heavy overhanging implements.

FOR DISCING AND CULTIVATING, you can control penetration exactly, even in changing soils.

FOR UTILITY JOBS, you can raise, lower, hold, and operate implements while you're working.

Quite a good explanation of those two new hydraulic levers. February 1956.

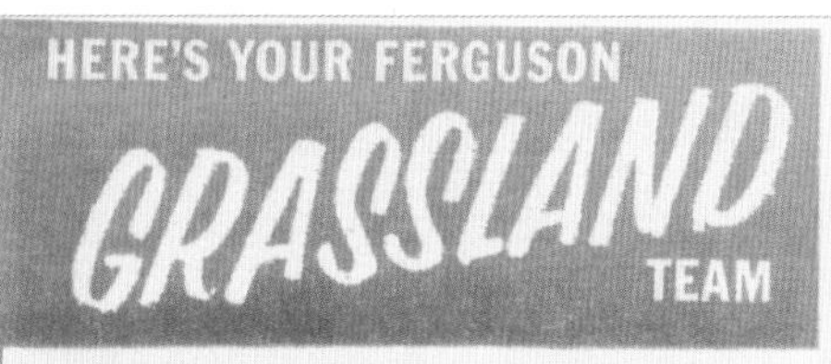

FASTEST MOWER EVER BUILT

• Dyna-Balance Drive • No Pitman
• Lead and Register Factory Set

A revolutionary new, high-speed mower with ground-level drive. Eliminates excessive vibration and resultant costly breakdowns so common with old-style, pitman-type mowers. Mow safely up to 30% faster because there's no vibration to tear your mower apart. Easier on the tractor and you, too!

FINEST RAKE ON THE MARKET

• Fully Mounted • 6-Bar Design
• 8 Foot • Double Bearings

Rake safely . . . with no leaf shattering . . . with no roping of windrows with a Ferguson PTO Rake. Side-delivers hay from swath to windrow in half the distance of ordinary rakes . . . and with a new Ferguson Tractor, the Rake turns at the exact, correct speed regardless of how fast you're traveling. Also available in 7-ft. size.

EASIEST BALER TO OPERATE

• Fast Maneuvering • Look-Ahead Baling
• Side-Mounted • "Self-Propelled"

With the Tractor-Mate Baler you turn faster on the headlands . . . get around easier in small or irregular fields. Baling is done right at your elbow. If a miss should occur, you see it before it leaves the chamber. Twin feeder fingers prevent plugging, and minimize leaf loss. Hooks up in less than 2 minutes.

HANDIEST FORAGE HARVESTER

• Row Crop • Direct Cut • Windrow Pickup
• Side-Mounted • "Self-Propelled"

This years-ahead Ferguson Tractor-Mate Forage Harvester gives self-propelled efficiency, yet you have your tractor for other farm jobs. It's side-mounted for maximum maneuverability . . . eyes ahead visibility. Takes only 90 seconds to hitch or unhitch. Precision controlled from tractor seat. *Ferguson*, Racine, Wis.

ASK YOUR DEALER FOR A DEMONSTRATION

Ferguson

BALER TO OPERATE

• Side Mounted • "Self-Propelled"
• Fast Maneuvering • Look-Ahead Baling

With the new Ferguson Baler you turn faster on the headlands . . . get around easier in small or irregular fields. Baling is done right at your elbow where you can watch it without getting off course. If a miss should occur, you see it before it leaves the chamber. Twin feeder fingers prevent plugging and minimize leaf loss. Hooks up in less than 2 minutes. There's *nothing* like a Ferguson Tractor-Mate Baler. Ask for a demonstration on your farm.

HANDIEST

FORAGE HARVESTER THERE IS

• Side Mounted • "Self-Propelled"
• Choice of 3 Heads • Powerful Cutter-Blower

This years-ahead Ferguson Forage Harvester gives self-propelled efficiency, yet you have your tractor for other farm jobs. It's side mounted for maximum maneuverability . . . eyes ahead visibility. Takes only 90 seconds to hitch or unhitch. Precision controlled from tractor seat. Your Ferguson Dealer can show you features galore that make *this* Harvester the best buy for your money. Ask him. *Ferguson*, Racine, Wis.

Ferguson

Ferguson designers made desperate attempts to take the little Ferguson tractor into the harvesting world with side mounted foragers, balers and prototype side mounted combine harvesters. But it was to be a short lived venture as Massey-Harris were already world leaders in the harvesting field.

How Ferguson Variable-Drive PTO Provide Both Ground Speed and Engine Speed Drive

At last! *One* tractor that lets you operate PTO-driven implements at either *ground speed* or *engine speed*, depending upon the need.

Simply move the lever to "Ground" PTO and you're ready to do a perfect raking job . . . distribute seed or fertilizer evenly in direct ratio to ground covered—because the PTO shaft will always revolve the same number of times per foot of forward travel, regardless of tractor speed.

Or you can select "Engine" PTO and the shaft rotates in ratio to *engine* speed for such jobs as harvesting, mowing, belt work. It's that simple. *It's that convenient!*

And, in addition, you get those other 4-Way Work Control features that have come to mean so much to Ferguson owners: Quadramatic Control, Dual-Range Transmission and "2-Stage" Clutching—all designed to help you farm more, work less.

Ferguson Dealers have had *years* of experience in sales and service of the *original* Ferguson System. Call your local Dealer today for a demonstration of the Ferguson "35" . . . and *feel the difference* on your day-to-day jobs. *Ferguson*, Racine, Wisconsin.

Ferguson

FOR HARVESTING you'll use the power take-off that's driven directly from the engine.

SEEDING-FERTILIZING utilizes "ground" drive for even distribution at uniform rate.

FOR RAKING put the PTO shift in "ground" position for rake action in ratio to forward travel.

Explaining the benefits of the variable drive PTO 1956.

The FERGUSON 35

Plows more acres per day ... easier and at lower cost

Get the tractor that saves time and money

Now that fall plowing time is here the man who owns a Ferguson 35 is lucky. He'll get the job done faster. He knows from experience that Quadramatic implement control, the Ferguson System's superior lugging power and matchless handling ease will put him acres ahead of his neighbors every day—unless they drive "35's", too.

On top of all this he knows the powerful Ferguson 35 engine saves him money. He actually does all this extra work at no increase in his fuel bill.

Now, don't just envy the man on a 35 . . . get one yourself and enjoy the work-saving, money-saving difference over any other tractor. Your Ferguson dealer will give you the full details . . . test drive a 35 and you'll find there is nothing like it.

Ferguson Tandem Disc Harrow, 3-point mounted. Discs up to 33 acres in 10 hours. Angle and depth easily adjusted from tractor seat. Choice of 16 or 18-inch discs, heat-treated steel. Transports conveniently in raised position.

ASK YOUR DEALER FOR A DEMONSTRATION

Ferguson TILLER. Check erosion, water run-off, weed growth—prepare seed beds with this exclusive Ferguson tool. Spring loaded teeth protect against damage from stones or roots. Excellent for renewing pastures or deep tillage work.

This advertisement is appearing in the current issue of the new Massey Harris Ferguson monthly paper "Farming Today."

CANADIAN FARM IMPLEMENTS—*October, 1956* 41

33 acres per day of discing – surely not if as shown the driver overlaps the work by 50 percent at each pass? October 1956.

February 2, 1957 THE FARMER

Which tractor stays at the barn?

One thing's for sure. When you own a Ferguson, it won't be the tractor you leave behind.

It'll be your favorite . . . on light jobs or heavy. On either, there's no better tractor for easy handling and low operating costs. Because the Ferguson is recognized as the one tractor with a completely integrated hydraulic system. No tedious, separate adjustments. You get full use of all your power—automatically.

When you're plowing, for instance: You can plow with 3 bottoms in most soils . . . at a cost the bigger, fuel-wasting tractors can't match. And you get this same power and economy on job after job.

Farmers often sum it all up by saying: "It's got the Ferguson System."

It's the System that gives you complete control. You can raise or lower implements . . . hold them in any position . . . maintain effective draft control . . . even change the speed of response.

That's how the Ferguson saves your job time and cuts operating costs. That's why we think you'll pick it for just about every tractor job. Why not see your Ferguson dealer? Talk performance. Talk trade. See him first . . . or last. Either way you can't lose. *Ferguson, Racine, Wis.*

SEE YOUR FERGUSON DEALER

see him **FIRST...** if you want the best deal but don't have time to shop around.

or see him **LAST...** if you still want to shop around and convince yourself.

EITHER WAY YOU CAN'T LOSE

POWERED TO SAVE YOUR JOB TIME

1957 marked the last year of the Massey-Harris-Ferguson period. This farmer seems well pleased with the progress of Ferguson farming to date! Subsequently the Ferguson badge was retained on the front of the bonnet, but those on the sides replaced with decals. Ferguson badging lasted well into the Massey Ferguson era in the USA unlike UK where it was dropped very early in the MF era.

THE FERGUSON FE35

The UK built FE35 tractor was the same as the USA built TO35 mechanically but with different livery and tinwork. It retained a diversity of models like the TE tractor series but not quite such a diverse range. Unlike its American counterpart it only ever had one colour scheme which was grey and gold. It came to be affectionately known as the "grey-gold." Another major difference to the TO35 was that the vast majority of tractors had diesel engines, only a few having 87mm petrol, tvo or lamp oil engines. The UK was always ahead of North America in adopting diesel power. The Standard 23C diesel engine had a notoriously bad reputation for poor starting, some of it perhaps well deserved, some of it undoubtedly down to operator inexperience with diesels. All FE35s had 12V ignitions – again the UK was quicker to adopt 12V than North America. The FE35 was a usefully more powerful and heavier tractor than the TE20. The 87mm petrol engine was a very smooth unit offering plenty of torque – more than the diesel.

74,655 "grey-golds" had been built by late 1957 at which point it went into red and grey Massey Ferguson livery as the MF35. This tractor really came into its own when MF replaced the Standard 23C diesel engine with a Perkins A3-152 three cylinder engine in 1959. This completely revolutionised the tractor and it became one of the "tractor greats".

Some 9000 Ferguson 35s were made in 1956 succeeding the production of the Ferguson TE20 which ceased in the same year. This is an early double page advertisement announcing the new Ferguson 35 which became commonly known as the "grey gold" Ferguson because of its livery. October 1956.

now

Twenty years ago— remember?—Ferguson *transformed* farm mechanisation. That first Ferguson was so good that it revolutionised tractor design. So good that over a million Ferguson tractors have been sold.

a revolutionary advance in tractor design

the new

FERGUSON engineers might have been satisfied. They weren't. They watched you working — and used *your* experience to build a new, still better Ferguson. It has what no other tractor has . . . the all-important basic benefits of the Ferguson System . . . tried, tested, and now *improved* . . . plus a host of entirely *new* features.

You'll see them on these pages. Study them—and see for yourself how the Ferguson "35" outdates every other tractor!

FERGUSON

Faster, more accurate work!

Just select your working depth, and away you go! New Ferguson System maintains that depth automatically and accurately. No need for you to make hurried adjustments. No need to worry about hidden obstacles — you've the new Ferguson overload release. You just work — faster and more evenly than ever before! *You get more output — in less man-hours.*

more output – less man hours = a greater margin of PROFIT with the **FERGUSON** 35

A joy to handle!

New, better steering lets you turn the new Ferguson 'on a sixpence'. It damps out 'kick' too, makes the operation lighter, smoother. You work both independent brake pedals with the *right* foot only: your left foot is completely free to operate the new dual clutch. Think how this helps you to have more done faster and better! *You get more output — in less man-hours.*

Make no mistake—the Ferguson 35 is the biggest advance in farm mechanisation in twenty years. Try it yourself. *Send the coupon today.* Test the Ferguson 35 *now*. You'll see how it's years ahead. You'll see how it gives greater output—in less man-hours—on *your farm*. You'll see how you raise your margin of PROFIT—with the Ferguson 35!

Two Power-take-offs in One Live P.T.O. Ground Speed P.T.O.

In the new FERGUSON 35 you've the choice of two different P.T.O. Drives. You can select a drive that's in ratio to the engine speed for mowing, baling, combining, etc., or you can select a drive that's in ratio to the ground speed of the tractor for such jobs as hayraking, planting, fertilising, weed controlling, etc., A P.T.O. Drive to suit any job.

Whatever you're doing — slow work or fast, one of the Ferguson six speeds gives you the speed you require, from 0.3 m.p.h. to 14 m.p.h.

You can save time and fuel — and the work is done better! *You get more output — in less man hours.*

FREE DEMONSTRATION — POST NOW!

To : Massey-Harris-Ferguson (Sales) Ltd., Coventry, Warwickshire.
Please arrange a free demonstration of the Ferguson 35 on my farm as soon as possible.

NAME ....

ADDRESS ....

Ferguson tractors, sold by Massey-Harris-Ferguson (Sales) Ltd., are manufactured by The Standard Motor Co. Ltd., Coventry

GO YEARS AHEAD — GET A

FERGUSON 35

outdates every other tractor!

FERGUSON 35

Never before a tractor with all the features you need!

Look carefully at these new Ferguson features. Compare the Ferguson 35 with all other tractors. You'll see how it outdates them — makes them old-fashioned already. And remember this: *only* Ferguson has what you *must have* for modern, efficient farming — the world-proved Ferguson System.

See what the new Ferguson does for *you*. Read how it does more — quicker and better than ever before. How it increases output — cuts man-hours. How you raise your margin of PROFIT — with the Ferguson 35.

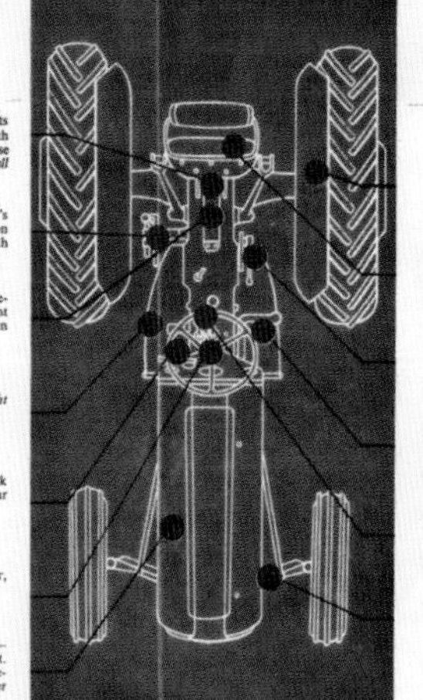

NEW — Hydraulic control! ★ Raise and lower implements ★ Select and maintain uniform working depth ★ Adjust hydraulic system's speed of response ★ Hold implements rigidly in any position *all without moving your hand six inches!*

NEW — 'Two-way' depth control! Controls implement's depth whether top link is in compression *or* tension . . . gives 'two-way' *automatic* depth control with even greater accuracy in working!

NEW — Constant overload release! Operates at pre-determined overload whether using heavy or light tillage implements — gives perfect operation in work and protection to implements.

NEW — Independent brake pedals! Both worked by *right* foot, leave left foot free to operate dual clutch.

NEW — Extra-safe single switch starter! *Cannot* work unless gear lever is in neutral — starting-in-gear is impossible!

NEW — Steering! Reduces 'kick,' makes operation lighter, smoother.

NEW — Great Diesel engine, 37 Brake Horse Power! — Power and power in reserve whenever you want it. A high-torque, high-compression, higher horse-power engine — that keeps you going when other tractors can't!

NEW — Brake shoes for greater safety, simplified for easier adjustment!

NEW — Comfort and convenience features! Include new, more comfortable seat and foot-rests, new removable radiator grille, new instruments, new servicing panel, and new larger fuel-tank with spill-over channel.

NEW — P.T.O. *selection*! Simple P.T.O. change lever selects *either* P.T.O. drive direct from the engine, or P.T.O. drive from the rear axle. Have the P.T.O. shaft revolving proportional to the engine speed or proportional to the ground covered.

NEW — Dual clutch! Control *both* tractor transmission *and* P.T.O. with left foot on a single clutch pedal. Press pedal half-way down to disengage transmission, while P.T.O. and pump continue to work. Press all the way down to disengage transmission *and* P.T.O.

NEW — Gear box! Six forward speeds and two reverse — a gear and a speed for every job on the farm. 'Creep' as slow as 0.3 m.p.h. — or speed up to 14 m.p.h. for fast jobs!

NEW — Axles! Heavy duty front axle mounting. Robust new rear axle for heavy loading — fitted with transmission-overload safety feature.

***AND ALSO NEW* 'Live' P.T.O. drive. For better, quicker control of mowers, balers, combines, other P.T.O. and hydraulic equipment.**

This five-page advertisement was the biggest found in the research for this book. The tractor certainly needed pushing on the market, as the diesel engine version – the most common – soon gained a poor reputation because of its poor starting. And yet some specimens were good starters. Late 1956.

Farm Implement and Machinery Review.—Feb. 1, 1957. 1547

One-man drilling and fertilising!

At your Ferguson Dealer's, now, there's the equipment you need for better, more economical drilling and fertilising. With the Ferguson Multi-Purpose Seed Drill and Fertiliser Attachment *one man* gets the work done—quicker, more accurately and at less cost than ever before!

FERGUSON MULTI-PURPOSE SEED DRILL, for use with "35" or TE-20 tractor: * sows a greater range of seeds than any other drill * single-speed drive, no gear changing * accurate sowing from 7 bushels cereals down to 2 lbs. turnips or swedes per acre * 9 ft. wide overall, yet sows up to 13 rows * low overall height for easy filling of large capacity hopper.

FERTILISER ATTACHMENT: * easily cleaned in minutes * hopper-taps corrosion-resistant * completely separate feed unit mechanism * uniform fertiliser flow, rates 110—1,200 lbs. per acre * self-lubricating gears and bearings.

See the SEED DRILL & FERTILIZER

you need - at your

FERGUSON

dealer's now

Ferguson tractors, sold by Massey-Harris-Ferguson Ltd., are manufactured by The Standard Motor Co. Ltd., Coventry.

H

Back to the well espoused TE20 theme of only one man on a job. Look at that tremendously wide setting of this Ferguson 35's front axle to facilitate easy following of the drill mark. February 1957.

February, 1957 FARM MECHANIZATION 27

NEW FERGUSON FE.35 LOADER COMPLETES YOUR FARMYARD MANURE-HANDLING TEAM...

for fastest-ever, one-man muck-shifting

NEW FERGUSON LOADER • Attached to Ferguson 35 tractor, and detached, in minutes • Lifts 17½ cwt. to 9′ 11″ • Compact design allows easy manoeuvring in confined spaces • Operator mounts tractor easily from either side, works Loader without leaving tractor seat • Fork returns automatically to loading position after dumping • Also available with hydraulic tip-off bucket.

This is the most important advance in manure handling ever offered to the British farmer. With the new Ferguson Loader, quickly and simply attached to the new Ferguson tractor and the Ferguson Spreader, one man gets the work done more efficiently and more economically than ever before.

Above are some of the important advantages of the new Ferguson Loader. But the *vital* advantage is what they all add up to. *More output — in less man-hours.* A greater margin of PROFIT. Prove this for yourself with a free demonstration on your own farm. *See your Ferguson Dealer — today.*

FERGUSON SPREADER • Hitched from tractor seat • Spreading rates from 4 up to 16 loads per acre.

GO AHEAD-GO FERGUSON

Ferguson tractors, sold by Massey-Harris-Ferguson Ltd., are manufactured by The Standard Motor Co. Ltd., Coventry.

EI

A stronger tractor needed a stronger loader. The 35 loader was a considerable design improvement on those offered for the TE20, and easier to operate. February 1957.

Farm Implement and Machinery Review.—Apr. 1, 1957. 1821

Ridging . . . Planting . . . Fertilising . . . Covering . . .

FOUR PLANTING JOBS ARE *ONE* FERGUSON OPERATION!

You save time and money when you plant potatoes the Ferguson way. You team up the Ferguson Ridger, Potato Planter and Fertiliser Attachment with the new Ferguson 35 tractor, or your present TE-20 model, in a matter of minutes . . . get your ridging, planting, fertilising and covering done in one fast, easy operation! When you come to harvest time, the Ferguson Potato Spinner helps you get the crop in quicker than ever before!

FERGUSON RIDGER:

* quickly mounted on tractor by 3-point linkage * working depth finger-tip controlled from tractor seat * kept in perfect alignment with tractor front wheels by unique Ferguson steerage fin * row spacings from 24"-30".

FERGUSON POTATO PLANTER:

* easily mounted on Ridger frame * plants 2 rows at a time —spacing from 8"-20" between plants * land-wheel drive operates bell indicator, ensures accurate seed spacing.

FERTILISER ATTACHMENT:

* secured above tractor rear axle by simple 3-point mounting * ensures steady, uniform fertiliser flow, at one of ten selected rates * feed mechanism controlled from tractor seat.

See the POTATO PLANTING equipment you need - at your **FERGUSON** dealer's <u>now</u>

Ferguson tractors, sold by Massey-Harris-Ferguson Ltd., are manufactured by The Standard Motor Co. Ltd., Coventry.

More sophistication of operations with the Ferguson 35 tractor era. Application of fertiliser at the same time as planting but in dry, windy weather the dust from the wheels on to the planters could be irksome. April 1957.

194 *Farm Implement and Machinery Review.—June 1, 1957.*

GRAIN'S IN THE GRANARY FASTER WITH FERGUSON!

Team up a Ferguson 3-ton trailer with the new Ferguson 35, or the TE.20, and get your grain home in double quick time! With Ferguson's exclusive automatic hitch, the trailer is coupled and uncoupled in minutes — and its sturdy construction and powerful brakes enable you to shift the heaviest loads in complete safety. What's more, the trailer can be tipped by hydraulic control to 40° in less than a minute!

See the 3-TON TRAILER you need - at your **FERGUSON** dealer's <u>now</u>

Ferguson implements fit *all* Ferguson tractors.

Grain in the granary with Ferguson and Massey-Harris. Ferguson never managed to get their side-mounted prototype combines into production. Further combine development was stopped as a result of the merger with Massey-Harris. June 1957.

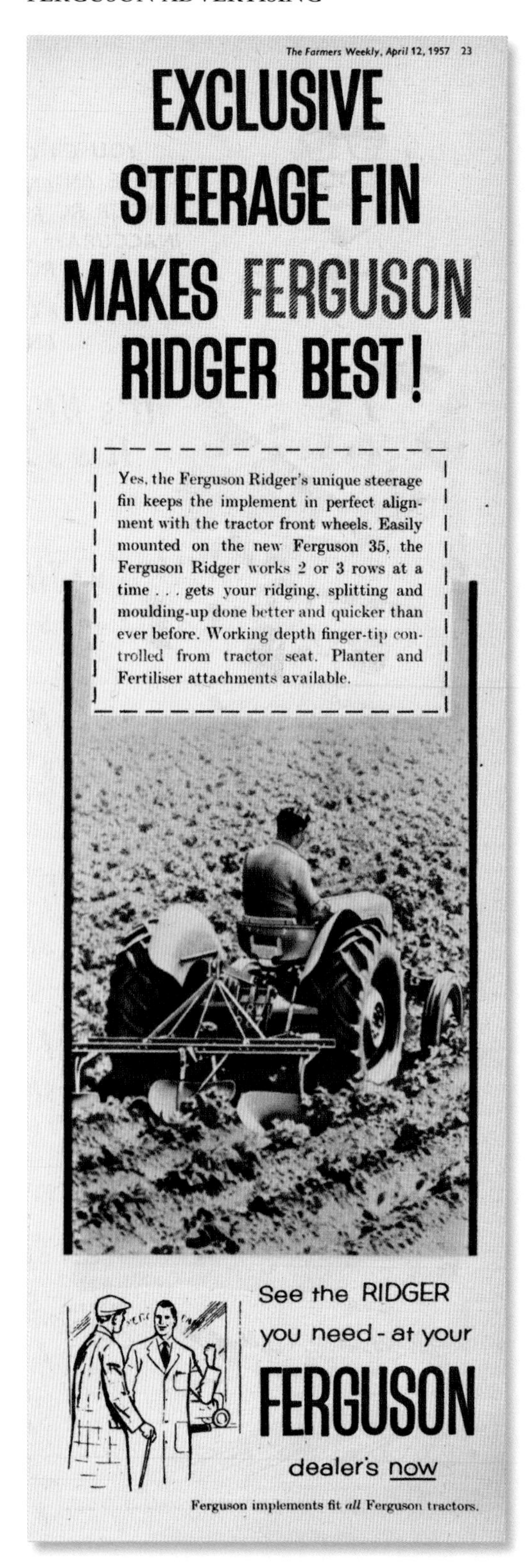

The Farmers Weekly, April 12, 1957 23
EXCLUSIVE
STEERAGE FIN
MAKES FERGUSON
RIDGER BEST!
Yes, the Ferguson Ridger's unique steerage fin keeps the implement in perfect alignment with the tractor front wheels. Easily mounted on the new Ferguson 35, the Ferguson Ridger works 2 or 3 rows at a time . . . gets your ridging, splitting and moulding-up done better and quicker than ever before. Working depth finger-tip controlled from tractor seat. Planter and Fertiliser attachments available.
See the RIDGER
you need - at your
FERGUSON
dealer's now
Ferguson implements fit all Ferguson tractors.

Blink your eyes and this shot could hark back to Ferguson Brown days. No improvement in standard of operation, just the speed of operation with more power. April 1957.

The Farmers Weekly, May 10, 1957 25
Exclusive automatic hitch gives...
QUICKER
TURN ROUND WITH
FERGUSON TRAILERS
Yes, Ferguson's exclusive automatic hitch lets you couple and uncouple the Ferguson 3-ton trailer in minutes—without leaving the seat of your new Ferguson 35 or TE-20 tractor. Think how this saves labour—and precious time! What's more, the Ferguson trailer's sturdy construction and the Hay Lades and Platform Extension accessories enable you to shift heavy, large-volume loads in complete safety. Manoeuvring's easy and, because one third of the weight of the load is transferred to the tractor, the trailer goes almost everywhere too!
See the TRAILER
you need - at your
FERGUSON
dealer's now
Ferguson implements fit all Ferguson tractors.

Not many original lades to be found now. They gave the Ferguson trailer great bale-carrying capacity. 1957.

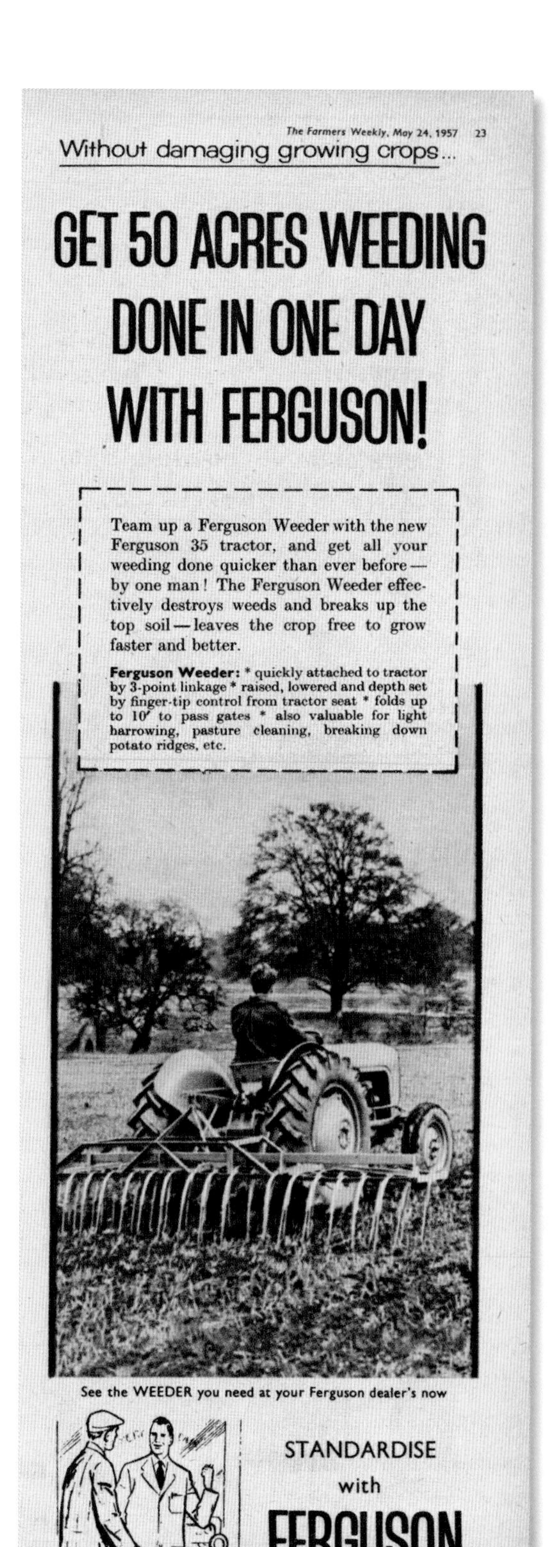

The Farmers Weekly, May 24, 1957 23
Without damaging growing crops...
GET 50 ACRES WEEDING
DONE IN ONE DAY
WITH FERGUSON!
Team up a Ferguson Weeder with the new Ferguson 35 tractor, and get all your weeding done quicker than ever before—by one man! The Ferguson Weeder effectively destroys weeds and breaks up the top soil—leaves the crop free to grow faster and better.
Ferguson Weeder: * quickly attached to tractor by 3-point linkage * raised, lowered and depth set by finger-tip control from tractor seat * folds up to 10' to pass gates * also valuable for light harrowing, pasture cleaning, breaking down potato ridges, etc.
See the WEEDER you need at your Ferguson dealer's now
STANDARDISE
with
FERGUSON
Ferguson implements fit all Ferguson tractors.

The 50 acres a day claim might seem far fetched but it was definitely possible because of its 13ft working width, low draft and operation at high speed. May 1957.

The Farmers Weekly, June 7, 1957 23

The new Ferguson 35 gives you...

"DUAL" CLUTCH

FOR TRANSMISSION AND P.T.O. CONTROL

Yes, on the Ferguson 35 you've a single clutch pedal which controls *both* tractor transmission and P.T.O:

Pedal half way down disengages transmission only—P.T.O. works on.

Pedal all the way down disengages both transmission and P.T.O. So the Ferguson 35 "dual" clutch allows you to work such machines as the Baler, Mower and Combine Harvester continuously, regardless of tractor starts and stops.

"Dual" clutch is just one more advantage of Ferguson farming—an advantage which only the Ferguson 35 gives you. See the Ferguson 35 on Stand 57 at the Royal Highland Show.

Dual clutch is available only on the De-Luxe Ferguson 35.

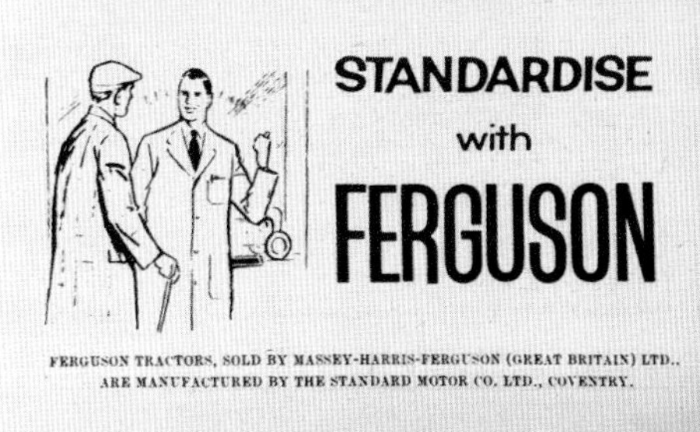

STANDARDISE with FERGUSON

FERGUSON TRACTORS, SOLD BY MASSEY-HARRIS-FERGUSON (GREAT BRITAIN) LTD., ARE MANUFACTURED BY THE STANDARD MOTOR CO. LTD., COVENTRY.

The dual clutch was a boon for pto operations and loader work. However, depending upon your height it could give you severe aching knees and was a heavier clutch to operate than the earlier Fergusons. June 1957.

The Farmers Weekly, June 7, 1957 25

The new Ferguson 35 gives you...

EIGHT SPEED GEARBOX- A SPEED FOR EVERY JOB!

You've six forward gears and two reverse on the new Ferguson 35. That means a speed for every kind of job on the farm—a speed at which you get the job done with *maximum* efficiency.

When you're transplanting, for instance, you can slow right down to 0.3 m.p.h.—the tractor still runs smoothly and you work more accurately. When you're shifting loads on the other hand, or simply moving from one job to another, the 35 speeds you along at 14 m.p.h.—saves you precious time and gets the work finished quicker!

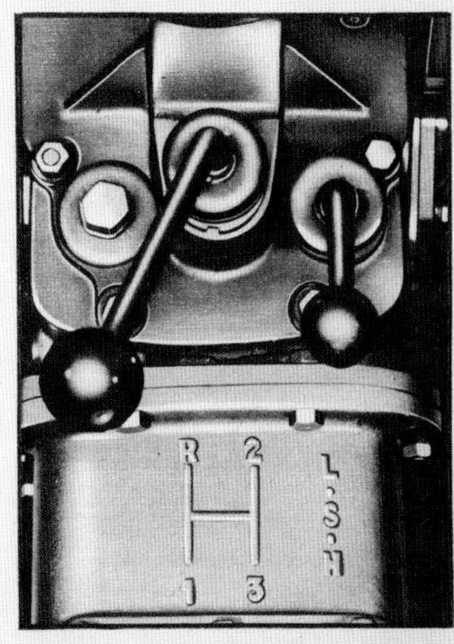

See the Ferguson 35 on Stand 57 at the Royal Highland Show

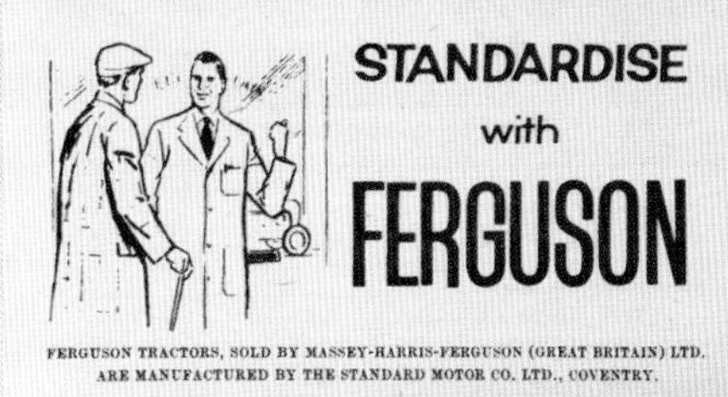

STANDARDISE with FERGUSON

FERGUSON TRACTORS, SOLD BY MASSEY-HARRIS-FERGUSON (GREAT BRITAIN) LTD. ARE MANUFACTURED BY THE STANDARD MOTOR CO. LTD., COVENTRY.

The extra low speeds available on the 35 tractor were very useful for such tasks as transplanting. However the lowest speed was not nearly as low as that offered by the Howard reduction gearbox which had often been fitted to TE20 tractors by market gardeners. June 1957.

The Farmers Weekly, July 19, 1957 21

NEW FERGUSON FE.35 LOADER COMPLETES YOUR FARMYARD MANURE-HANDLING TEAM

- for fastest-ever, one-man muck-shifting

New Ferguson Loader • Attached to Ferguson 35 tractor, and detached, in minutes • Lifts 17½ cwt. to over 9' 0" • Compact design allows easy manoeuvering in confined spaces • Operator mounts tractor easily from either side, works Loader without leaving tractor seat • Fork returns automatically to loading position after dumping.

This is the most important advance in manure handling ever offered to the British farmer. With the new Ferguson Loader, quickly and simply attached to the new Ferguson tractor and the Ferguson Spreader, one man gets the work done more efficiently and more economically than ever before.

The new Ferguson Loader—with many other Ferguson implements you need—is at your Ferguson Dealer's *now*. Ask him for a free demonstration on your own farm, *today*.

See the Loader you need at your Ferguson dealers now

STANDARDISE with FERGUSON

Ferguson 35 loader on Ferguson 35 tractor loading a Massey-Harris 712 manure spreader. A good team but overloading of the land drive spreader could cause it to jam in wet conditions. It was then an arduous task to unload the muck from behind the jammed beaters. July 1957.

July 1957 PRACTICAL POWER FARMING 61

Exclusive automatic hitch gives...

QUICKER TURN ROUND WITH FERGUSON TRAILERS

Yes, Ferguson's exclusive automatic hitch lets you couple and uncouple the Ferguson 3-ton Trailer in minutes—without leaving the seat of your new Ferguson 35 or TE-20 tractor. Think how this saves labour—and precious time! What's more, the Ferguson Trailer's sturdy construction enables you to shift heavy loads in complete safety. Manoeuvering's easy and, because one third of the weight of the load is transferred to the tractor, you get better traction in bad conditions. 30 cwt. Trailer also available.

See the TRAILER you need at your Ferguson Dealer's now.

STANDARDISE with FERGUSON

A fair weight on the trailer but no sign of the tractor rearing up with the Ferguson pick up hitch which placed the weight of the trailer beneath the rear axle, and the weight transfer achieved means that pulling up the slope is no problem. July 1957.

Let Ferguson design your plough around your needs. October 1957.

July, 1957 FARM MECHANIZATION 49

ONE-MAN TEAM FOR FASTER LOAD SHIFTING

Quickly and simply attached to the new Ferguson 35 tractor, the Ferguson 3-ton Trailer and new Ferguson F.E. 35 loader get *every* load on your farm shifted more economically than ever before. *One man* gets the work done—in *double* quick time!

Ferguson F.E. 35 Loader:
* Attached and detached in minutes * Lifts 17½ cwt. to 9′ 11″ * Compact design allows easy manoeuvering in confined spaces * Operator mounts tractor easily from either side, works Loader without leaving tractor seat * Available with manure fork or hydraulic controlled tip bucket.

Ferguson 3-ton Trailer:
* Coupled and uncoupled by exclusive Ferguson automatic hitch * Sturdy construction enables you to shift the heaviest loads in complete safety * Easily manoeuvered * One third of weight of load transferred to tractor—for better traction in bad conditions * 30 cwt. Trailer also available.

See this equipment at your Ferguson Dealer's now.

STANDARDISE with FERGUSON

The FE 35 loader could be fitted with a materials bucket as well as a muck fork. It could lift an astonishing 17.5cwt. July 1957.

FARMER & STOCK-BREEDER, 10 September 1957
25
"I get more done with one man—
I'm standardised with Ferguson"
Farmers who standardise with Ferguson enjoy many advantages—important advantages that save a great deal of money and man-hours.
For when you standardise with Ferguson, your tractor and implements are engineered to work together. This saves you fuel–because you work better and faster. It saves you time—because you change from one implement to another in minutes, and, since all Ferguson implements work on basically the same principle, you learn to operate a new implement quickly. And it saves you money, because planned safety devices cut costly repairs. To cap it all, you make a vital saving on labour—for nearly all Ferguson implements are operated by one man.
That is why it is well worth your while to standardise on Ferguson equipment. See some of this equipment on the following pages.
Standardise your farm machinery with Ferguson
for more food at less cost

Kids stuff. Anyone can see that a modern Ferguson beats an old Fordson type tractor, and Ferguson implements often saved a man. September 1957.

The Farmers Weekly, November 1, 1957 17
"Why do they give medals to tractors, Daddy?"
Six great honours have been won by the Ferguson 35 tractor since its introduction in October, 1956:
Silver Medal, Royal Highland Show, 1957
Silver Medal, Royal Ulster Show, 1957
Silver Medal, and Burke New Implement Trophy, "Royal" Show, 1957
Silver Medal, and Albon Davies Cup, Royal Welsh Show, 1957.
Five of these six awards have never before been won by a tractor. But, in the opinion of the experts, the Ferguson 35 represents such a remarkable advance in farming equipment that it merits these unusual awards.
The success of the tractor reflects the effectiveness of Ferguson Standardisation. For when you standardise with Ferguson, your tractor and implements are engineered to work together — to produce more food at less cost.
Be sure of success on your farm, and standardise on Ferguson equipment. See some of this equipment on the following pages.
Standardise your farm machinery with Ferguson
for more food at less cost

Six medals in a year. An amazing feat for a tractor that had the poor starting reputation! But in fairness the combination of gearbox, clutch and hydraulic control improvements was quite a package. November 1957.

May, 1957 FARM MECHANIZATION 27

WEED CONTROL IS A LOW-COST ONE-MAN JOB–WITH FERGUSON

Team up these Ferguson implements with a new Ferguson 35 tractor, or the TE-20 model, and get all your weeding done quicker and more economically than ever before — by one man!

FERGUSON WEEDER

• gets 50 acres weeding done in one day • quickly attached to the tractor by 3-point linkage • raised and lowered from tractor seat • effectively destroys weeds without damage to crop • folds to 10′ to pass gates.

FERGUSON MEDIUM PRESSURE SPRAYER

• application rates from 10-80 gallons per acre • quickly mounted on tractor linkage • easily operated and manoeuvered—spring-loading automatically protects booms against damage from obstructions • pressure range, mechanical agitation and piston type pump permit use of most modern weed-killers • quality built throughout • Low Volume Sprayer also available.

See the WEED KILLING equipment you need – at your **FERGUSON** dealer's now

Ferguson implements fit *all* Ferguson tractors.

FI

Chemical crop protection was starting to advance in the late 1950s, but there was still a place for the tried-and-tested high-output weeder. This type of mechanical weeder has made a comeback in recent years with the advent of organic farming methods. 1957.

18 The Farmers Weekly, November 1, 1957

FERGUSON TYRE TRACKS * *are fitted in minutes* * *allow track settings of 52", 56" and 60"* * *are flexible, but grip tightly* * *absorb minimum engine power.*
FERGUSON 3-TON TRAILER * *hitched and unhitched by driver from tractor seat* * *transfers one third of weight of load to tractor wheels* * *tipping model tips by fingertip control to 40° in less than a minute* * *non-tipping model also available.*

Keep going

WHEN THE GOING'S TOUGH!

At your Ferguson Dealer's, now, there's the equipment you need for the tough going you expect just now. With the new Ferguson Tyre Tracks you always get on with any job—through *all* conditions of weather. And the Ferguson 3-ton Trailer lets you take heavy loads through boggy land, wet clay, loose top soil, up steep slopes...anywhere!

See the TYRE TRACKS and TRAILERS you need - at your **FERGUSON** dealer's now

Ferguson tractors, sold by Massey-Harris- Ferguson Ltd., are manufactured by The Standard Motor Co. Ltd., Coventry.

What could match a Ferguson trailer, Ferguson pick up hitch, the weight transfer achieved and the massive traction from Ferguson half tracks? 1957.

The Farmers Weekly, January 17, 1958 17

NO OTHER TRACTOR

gives you so much in features — performance and value

THE MASSEY-FERGUSON

- Power to match your requirements — Low-cost V.O. or 38 B.H.P. Diesel Engine. A higher torque, higher-compression engine — keeps you going when other tractors can't!
- Dual Clutch! Control both tractor transmission and P.T.O. with left foot on a single clutch pedal, and stop tractor while P.T.O. and pump continue to work.
- Superb manoeuvrability with independent brakes — or master brake control.
- Patented heavy-duty front axle requires no track rod adjustment. Robust new rear axle — fitted with exclusive transmission-overload safety feature.
- With an eye to the future, implement mounting pads will keep your 35 in line with new Massey - Ferguson machines to come.
- A truly versatile tractor with amazing earning power. It's compact, handy and has a low centre of gravity for safe hillside work.

—and the Ferguson System Control features shown on the preceding page.

STANDARDISE WITH MASSEY-FERGUSON

and farm with real economy

Manufactured by the Standard Motor Co. Ltd. Coventry. for Massey-Harris-Ferguson (Gt. Britain) Ltd.

Identical tractor – just a different badge and change of livery. Late 1957 saw the advent of the MF35 with the change of company name and dropping of the two line policy of marketing Massey-Harris and Ferguson lines separately. Just a few of these early red and grey 35s had the old Ferguson badges on the front and sides of the bonnet. January 1958.

100% TRACTOR TEAM!

Two great Ferguson System tractors, the Massey-Ferguson 65 and highly successful 35 will handle all your work — will raise output from each tractor and operator to the highest possible level. Both tractors have traditional Ferguson System advantages — exceptionally high power, with low weight — accuracy, efficiency, safety and economy of working unsurpassed by any other tractor. No other tractors form a finer team, because the 65 and 35 are designed for complete standardisation — with identical controls and one range of "tailor-made" interchangeable implements. For more work every man hour, for greater return, every tractor hour — standardise with Massey-Ferguson.

MASSEY-FERGUSON 35

Choice of V.O., Petrol or Diesel engine. Compact, manœuvrable, low centre of gravity — it's all the modern farmer needs for features, performance and value. Ferguson System Hydraulics, dual clutch, double P.T.O. Six forward speeds from creep upwards.

MASSEY-FERGUSON 65

The over 50 h.p., 4-5 furrow tractor that weighs only 4,000 lbs. Great power and new features including disc brakes, differential lock, full power steering, lower links with hinged ball ends for easy attachment of heaviest implements. Category I or 2 implements. Ferguson System Hydraulics, dual clutch, double P.T.O. Six forward speeds and two reverse.

FOR MAXIMUM EARNING POWER PER TRACTOR

STANDARDISE WITH MASSEY-FERGUSON

Manufactured by the Standard Motor Co. Ltd. Coventry, for Massey-Ferguson (Great Britain) Ltd.

End of the Ferguson 35 only era – it acquires a big brother in the form of the MF65 tractor. June 1958.

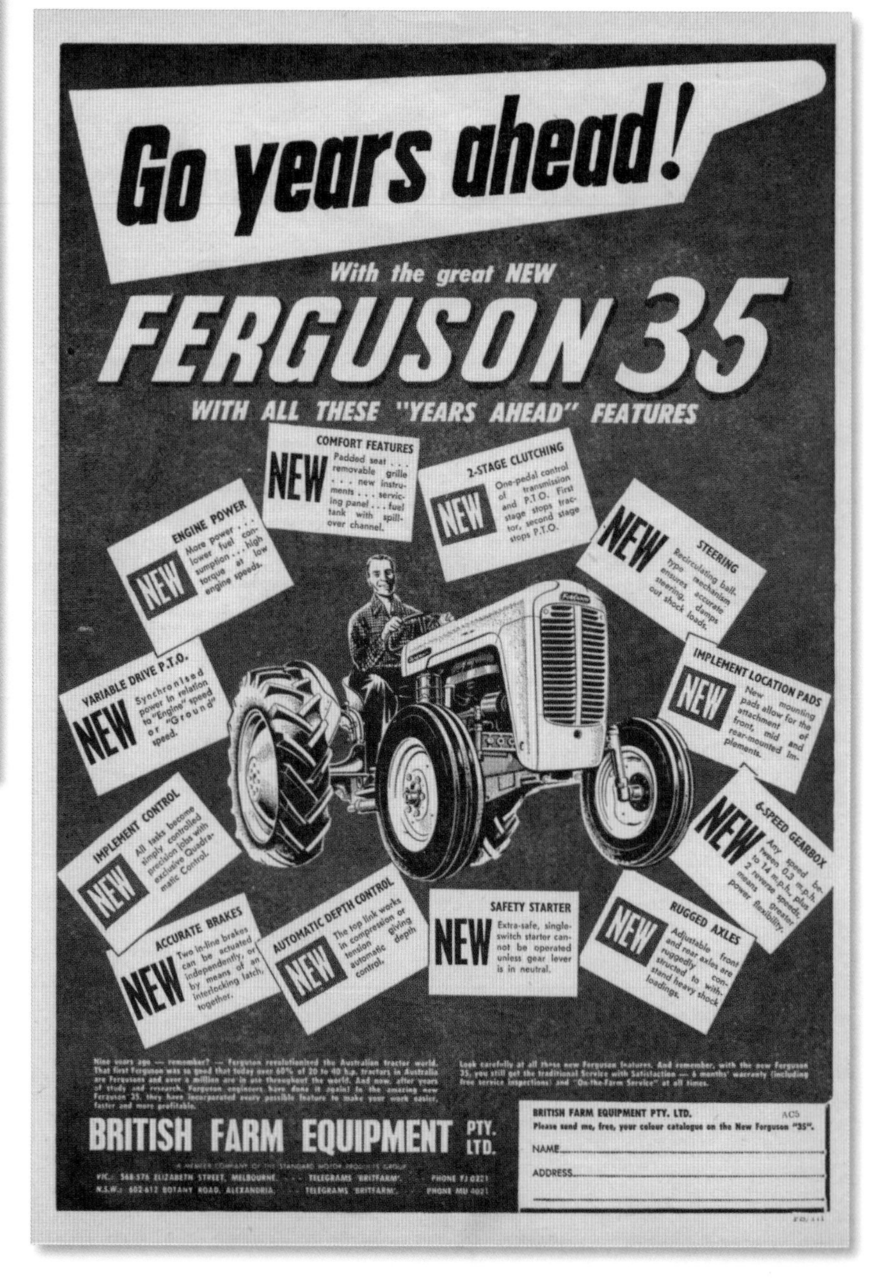

12 new features for the Ferguson 35 detailed by Australian Ferguson distributor, British Farm Equipment. May 1957.

IT'S HERE!
and we want YOU to drive it
THE SENSATIONAL NEW
FERGUSON '35'
NEW QUADRAMATIC CONTROL
This great new Ferguson exclusive lets you control every implement with the one compact control quadrant.
NEW VARIABLE DRIVE P.T.O.
Your choice of 2 different P.T.O. drives — engine speed or ground speed P.T.O.
NEW 2-STAGE CLUTCHING (LIVE P.T.O.)
You control both the tractor transmission and the P.T.O. with a single 2-stage clutch pedal.
NEW 6 forward speeds, 2 reverse
From 0.3 to 14 m.p.h. in the forward gears and from 0.4 to 6.8 m.p.h. in reverse.
HERE AT LAST! ALL THE FEATURES YOU HAVE EVER WANTED
-ALL IN ONE MIGHTY TRACTOR
NEW ENGINE POWER
Output is up—economy still amazingly low.
NEW AUTOMATIC DEPTH CONTROL
Depth is controlled whether top link is in compression or tension.
NEW RUGGED AXLES
Fitted with transmission overload safety feature.
NEW SAFETY BRAKES
For greater safety . . . easier adjustment and operation.
NEW STEERING
Reduces 'kick' and makes operation lighter and smoother.
NEW SAFETY STARTER
More comfortable seat and foot rests, new removable radiator grille.
NEW COMFORT FEATURES
Extra safe — cannot work unless gear lever is in neutral.
NEW IMPLEMENT LOCATION PADS
For front, rear or mid-mounted implements.
● HERE IT IS — the most advanced tractor in the world today — the New Ferguson 35. It only stands to reason that the world-wide experience gained all over the world by over 1,000,000 Ferguson owners would result in the best ever tractor being produced. Never before in one tractor have there been assembled so many features you must have for better, for easier and more profitable farming . . . and with this new Ferguson 35 you get famous Ferguson quality — the kind of construction that means years of efficient operation at the lowest possible upkeep. You get 6 months warranty and free 'On The Farm' Service, too — Ferguson still outvalues them all in price, performance and versatility. See the new Ferguson NOW at your nearest Ferguson dealer.
● Ferguson Tractors and Implements are distributed throughout Australia by . . .
VICTORIA
BRITISH FARM EQUIPMENT PTY. LTD.
568-576 ELIZABETH ST., MELB. FJ 0221
QUEENSLAND
BRITISH TRACTOR & IMPLEMENTS PTY. LTD.
156-158 SYDNEY ST., NEW FARM, LW 1011
WEST AUSTRALIA
BRITISH TRACTOR & MACHINERY CO.
1275 HAY ST., WEST PERTH. BA 2215
NEW SOUTH WALES
BRITISH FARM EQUIPMENT PTY. LTD.
602 BOTANY RD., ALEXANDRIA. MU 4021
STH. AUSTRALIA
BRITISH MECHANICAL FARMING LTD.
76 WAYMOUTH ST., ADELAIDE, S.A. LA1787
TASMANIA
BRITISH FARM MECHANISATION CO.
123 MURRAY ST. HOBART. TAS. B 2861
FERGUSON DEALERS, SERVICE AND SPARE PARTS EVERYWHERE THROUGHOUT AUSTRALIA
FS/125
102
April, 1957—POWER FARMING AND BETTER FARMING DIGEST
POWER FARMING AND BETTER FARMING DIGEST,—April, 1957
103

More detailing of the new features of the Ferguson 35 by Australian Ferguson dealers. April 1957.

The right price and big power claimed for the new Ferguson 35, available in six models including petrol, kerosene and diesel engines. November 1957.

Harry Ferguson always laid great emphasis on after sales service. February 1958.

Emphasising the unique 4 way control Ferguson hydraulic system available on all six models of the Ferguson 35. May 1958.

For a cost of from only £859 you could get an extensive range of features on a Ferguson 35 tractor. 1958.

The diesel Ferguson 35 was the principal engine type sold in the Ferguson 35s. It was claimed to be the diesel with a difference. Well, in retrospect they had a reputation for being poor starters! October 1959.

Mother and son bringing a tea break to father on his Ferguson 35 with side-mounted mower. Dad looks happy enough on his £859 bargain. January 1959.

A five-shank Ferguson spring-tine chisel plough soaking up the power from a Ferguson 35 – quite a load for a small tractor but it seems up to the job. April 1959.

Demonstrating the manoeuvrability of a Ferguson 35 in a small field enclosure. This demonstration was frequently used by Harry Ferguson back in the Ford Ferguson days. Being only slightly longer than the Ford Ferguson, the 35 is quite up to the task of not leaving any of the enclosure uncultivated. June 1959.

THE FERGUSON 40

The Ferguson 40 was the ultimate development of Ferguson System tractors whilst they stayed under the Ferguson badge. It was designed for and only sold in North America. It was introduced in 1956, its design predecessor the Massey-Harris 50 having been introduced in the previous year. The M-H 50 was designed in response to a clamour from M-H agents after the takeover of Ferguson by M-H for a Ferguson System tractor. The M-H 50 met this demand. In wanting a Ferguson System tractor the M-H dealers also wanted incorporated into it traditional North American tractor rowcrop capabilities. North America, with its vast areas of rowcrops, had traditionally relied on tricycle type tractors with mid-mounted cultivators. Harry Ferguson had always strenuously rejected anything approaching this type of design being incorporated into his tractors. Mid-mounted implements could not make use of his weight transfer principle. However, by now he was out of the recently formed Massey-Harris-Ferguson company. The resulting M-H 50/Ferguson 40 tractor would have been anathema to him! Although the tractor had mid-mounted implement capability it retained the conventional Ferguson three point, rear linkage of the TO35.

The M-H 50 and Ferguson 40 were identical mechanically differing only in the tinwork and livery. They basically comprised a TO35 rear axle/transmission/engine unit on to which were grafted four alternative chassis/wheel configurations – utility, single front rowcrop, V twin rowcrop and a hi-clear (hi-arch) version. All could have mid-mounted implements. In 1958 the M-H 50 and Ferguson 40 became the Massey Ferguson 50 with the change of company name, and was the same tractor as the M-H 50. Design wise these three tractors were the predecessors of the Massey Ferguson 65 tractors.

Some 9000 Ferguson 40 tractors were made in 1956 and 1957 alongside the Ferguson TO 35s. The advent of the Ferguson 40 was probably a response to the demand for high clearance row crop tractors which Ferguson had previously never produced. It is known that Harry Ferguson was very opposed to the idea of tricycle type tractors. This is a Canadian advertisement introducing the tractors. June 1956.

by Ferguson... in 3 New High-Clearance Models!

Ferguson Hi-40

WITH 4-WAY WORK CONTROL

If you prefer a high-clearance tractor and mid-mounted cultivators, be sure to see—and test drive—the great, new Ferguson Hi-40.

The Hi-40 has all the proved work-saving, money-saving features of its lower slung running mate, the famous Ferguson 35.

See These New Features

Both the Hi-40 and 35 now feature 12-volt electrical systems that get you off to faster, surer starts. You'll feel the surge of new, added power when you rev up the engine. And, if you like, there's power steering, too (factory installed as an integral part of the tractor).

Try 4-Way Work Control

The real thrill comes when you put the new Hi-40 to work. Control of implements was never easier, surer. The time-tested and continually improved Ferguson System gives you complete mastery over every farm job.

Right at your finger tips is new Quadramatic Control that will amaze you with its close command of implements. This Ferguson "first" lets you raise and lower implements, select draft and maintain working depth, adjust the hydraulic system's speed of response and hold implements in any position you choose—all with the same control quadrant.

As you sit in the comfortable "Foam-Float" seat with adjustable back rest, try out these other Ferguson controls: Variable-Drive PTO for versatile, synchronized power; "2-Stage" Clutching for one-pedal control of both transmission and PTO; and Dual-Range Transmission for peak efficiency in each speed range.

Ask for a Demonstration

Don't take our word for it. Test drive the Hi-40 on your farm. Your Ferguson Dealer will be glad to arrange a Spring demonstration for you. And Ferguson Dealers have had years of experience in sales and service of the Ferguson System. Only after such a demonstration will you fully realize why we say you'll farm more, farm better . . . at less cost, with a Ferguson. *Ferguson*, Racine, Wisconsin.

Ferguson

How Ferguson 4-WAY WORK CONTROL Gives You Complete Selectivity and Flexibility in Tractor Power and Implement Control

1. Quadramatic Control

At a touch of your finger tips you can *raise* and *lower* implements, select draft and maintain uniform working depth, adjust hydraulic system's speed of response, hold implements rigidly in any position selected.

2. Dual-Range Transmission

Wide range of speeds—6 forward and 2 reverse. You simply shift to the gear and speed that fits your work *exactly*. Convenient gear lever and standard automotive shifting make gear selection easy and natural.

3. Variable-Drive PTO

Shift to "Ground" speed and PTO is in direct ratio to forward movement of the tractor for raking, planting, fertilizing. Shift to "Engine" speed and PTO is in ratio to engine speed for forage harvesting, baling or mowing.

4. "2-Stage" Clutching

Single foot pedal allows you to control both transmission and live power take-off. You can operate machines like the Baler, Mower and Forage Harvester, continuously, regardless of tractor starts and stops.

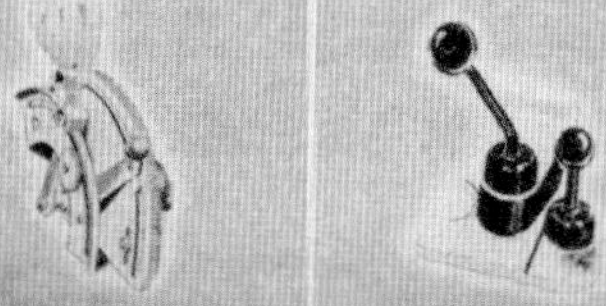

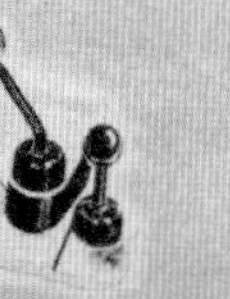

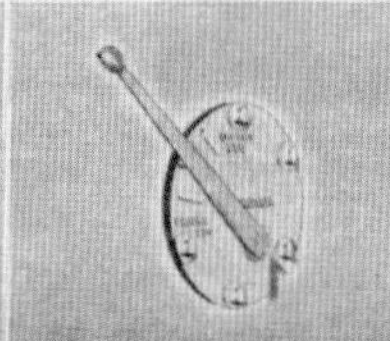

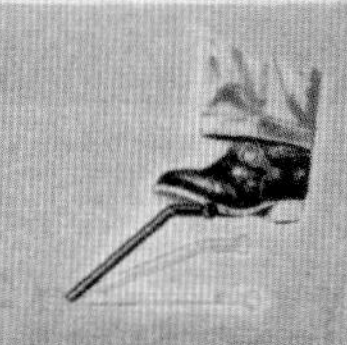

Stressing that the new Ferguson 40 has the same Ferguson advanced features as the famous Ferguson 35, but with the high clearance tricycle and four wheel row crop options.

Another Years-Ahead
TRACTOR BY
Ferguson
The Hi-40 with
Choice of 3
Front Wheel Styles
BRAND NEW
• QUADRAMATIC CONTROL
• "2-STAGE" CLUTCHING
• VARIABLE-DRIVE PTO
• DUAL-RANGE TRANSMISSION
• CONVERTIBLE FRONT WHEEL SYSTEM
• INCREASED POWER
• OPTIONAL POWER STEERING
• 12-VOLT ELECTRICAL SYSTEM
WIN FREE VACATION FOR 2
Exciting travel vacations to London and Paris . . . Hawaii . . . Caribbean Islands . . . Cuba . . . New York City. Ferguson Dealers have contest entry blanks. Nothing to buy or write. "Go places with Ferguson."
CHOICE OF MODELS. The Ferguson Hi-40 is available in models shown below. Front ends are also convertible—by the owner himself.
Four-Wheel Model
Dual-Wheel Tricycle
Single-Wheel Tricycle

Until the Ferguson 40 model, Ferguson had never offered mid mounted row crop equipment, yet for years this had been common with other row crop tractor makers.

DEALERS! ...here's the support you'd expect from a leader

FAIR PRICING of quality products provides lasting owner satisfaction.

EXPANDED LINE of advanced-design Ferguson Tractors and Implements.

MERCHANDISING SUPPORT in the form of sales tools and promotions.

ADVERTISING SUPPORT in national, state and specialized publications.

PRODUCT TRAINING to provide the foundation of good salesmanship.

SERVICE TRAINING SCHOOLS and aggressive parts programs.

FLEXIBLE FINANCING ASSISTANCE to supplement local credit.

A GREAT, NEW LINE OF FERGUSON TRACTORS

The latest in a procession of new and improved farm equipment, designed to meet the needs of mechanized agriculture, comes from Ferguson. And it reaches the farmer through successful Ferguson Dealers. If you are interested in the profit opportunity offered by a Ferguson Franchise . . . write or wire Ferguson, Racine, Wisconsin.

Ferguson

The Ferguson 40 was the same power as the Ferguson 35. But look at the different clearance.

Your farm is our showroom

A Ferguson looks its best with dirt on the tires. And there's no place like your own farm to size up its beauty.

That's where to see Ferguson precision control. In your field, cultivating 4 rows of corn that's 5 inches tall. Or 3 feet tall. You have plenty of clearance with a Hi-40.

The minute you get in the seat, you'll enjoy this tractor. Safe, comfortable, fun to operate. No tedious, separate adjustments. And no waste of fuel. And remember, the Ferguson is recognized as the one tractor with a completely integrated hydraulic system.

Lugging ability? Make your own field the test. Pick out a tough field. Hook up a Ferguson 3-bottom plow, and pull right through it. Automatic weight transfer and true draft control apply your power where it's needed . . . you get full use of all your power at a cost the bigger, fuel-wasting tractors can't match.

The proof is in the field—with your choice of five models. Ask your dealer to show you. In actions, not words. Talk trade with him. See him first . . . or see him last. Either way you can't lose. *Ferguson, Racine, Wis.*

SEE YOUR FERGUSON DEALER

see him **FIRST...** if you want the best deal but don't have time to shop around;

or see him **LAST...** if you still want to shop around and convince yourself.

EITHER WAY YOU CAN'T LOSE

POWERED TO SAVE YOUR JOB TIME

This photo really emphasises the clearance of the Ferguson 40. The lad is undoubtedly impressed! But could the tractor really deal with 3 feet high crops?

Ferguson
40
POWER
COMMAND
IT'S HERE IN THE
ADVANCED FERGUSON SYSTEM
40
Power Command is the integration of the advanced Ferguson System with other Ferguson firsts to give you complete and instant command over tractor and implement operation.
From your command post in the comfortable Foam-Float seat, with your finger tips on the new Quadramatic Control, you direct operations: raise and lower implements; select draft and maintain working depth; adjust the hydraulic system's speed of response; and hold implements in any position.
Contributing to your command over every farm job are other famous Ferguson 4-Way Work Control features: Variable-Drive PTO for versatile, synchronized power; "2-Stage" Clutching for one-pedal control of both transmission and live PTO; and Dual-Range Transmission for peak efficiency in every speed range.
Now, the one tractor that dared to be different years ago, still leads the way with five new models, every one of them with the famous Ferguson System: the new "40"; 3 "Hi-40" models with convertible front-wheel assemblies . . . and a better than ever Ferguson "35", proved on tens of thousands of farms.
Ferguson, Racine, Wisconsin.
NEW POWER STEERING
Even more valuable on your tractor than on your car. (Factory installed option.)
HEAVIER, LONGER, MID-MOUNT, TOO!
New Mid-Mounted Cultivators available for the Ferguson "40" and "Hi-40" models.
NEW, 12-VOLT ELECTRICAL SYSTEM
Faster, surer cold weather starts . . . hotter spark for better performance.
FREE VACATION TRAVEL CONTEST!
It's easy. Nothing to buy or write. See your Ferguson Dealer for entry blank.

Power at your command. A bigger tractor but only the same power as the Ferguson 35. But it's how you can use the power that really counts.

The following six advertisements were not traced from newspapers or magazines, but from a "Ferguson Newspaper Advertising brochure" that was sent to dealers. Dealers could then choose which they wished to place in local newspapers and ask Ferguson for the original "newspaper mats" with which to place the advertisement. April 1956.

BRAND NEW by FERGUSON

High-Clearance Hi-40 Handles Row Crop Cultivation and ALL Other Farm Chores With Greater Ease!

Now, the famous Ferguson System is available in three high-clearance tractor models which also feature new, added power—12-volt electrical system and optional power steering.

Ease into the comfortable "Foam Float" seat and test this years-ahead tractor. Right at your finger tips is new Quadramatic Control that will amaze you with its close command of implements. This Ferguson "first" lets you raise and lower implements, select draft and maintain working depth, adjust the hydraulic system's speed of response and hold implements at any position you choose—all with the same compact control quadrant.

And you'll find the other members of the Ferguson 4-Way Work Control team contributing their part to your mastery over every farm job: Variable-Drive PTO . . . "2-Stage" Clutching . . . Dual-Range Transmission.

Come in and see this new tractor. We'll let you test it, fitted out with mid-mounted cultivators. It's an experience you'll never forget.

DEALER'S NAME AND ADDRESS

Available in models shown below. Front ends are convertible, and by the owner himself

Four-Wheel Model

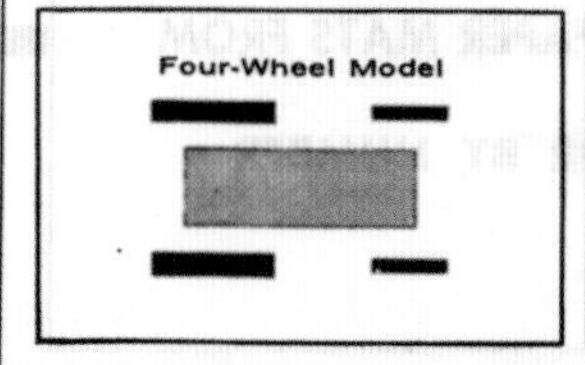

Dual-Wheel Tricycle

Single-Wheel Tricycle

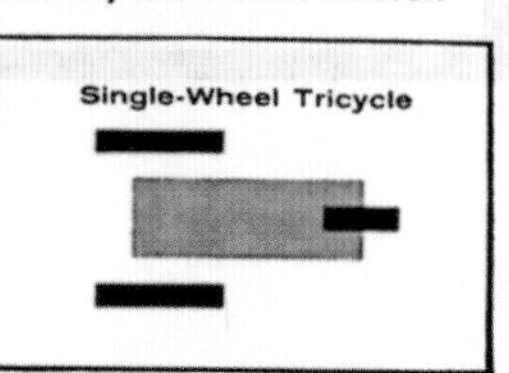

NOW TRY
The Sensational New
Ferguson 40
ADVANCED
FERGUSON SYSTEM
GIVES YOU COMPLET
POWER COMMAND
over both tractor and implement
More power, more positive control are now yours with a new Ferguson "40". You'll farm better, easier and with greater efficiency. The "40" is longer, heavier and can be used for mid-mount cultivation. Power steering and power wheels, too, if you wish. Also three high-clearance models. Remember: every Ferguson Tractor has the Ferguson System.
DEALER'S NAME
ADDRESS
PHONE
COME IN...WE'LL GIVE YOU A DEMONSTRATION

IT'S HERE NOW
THE TRACTOR SENSATION OF THE YEAR
NEW
Ferguson
40
LONGER, HEAVIER, MORE POWERFUL
WITH ADVANCED FERGUSON SYSTEM
Experience Complete
POWER COMMAND
for Yourself
More power, more positive command over tractor and implement operation are now yours with the beautiful new Ferguson "40". You'll farm better, easier and with greater efficiency. Can be used for mid-mount cultivation. Power steering, too, if you wish. Also 3 high-clearance models. Remember: every Ferguson Tractor has the Ferguson System.
DEALER'S NAME
ADDRESS
PHONE
COME IN...WE'LL GIVE YOU A DEMONSTRATION

You've read about it... NOW
PHONE FOR A DEMONSTRATION
TRY THE NEW AND POWERFUL
Ferguson 40
with ADVANCED FERGUSON SYSTEM
Experience Complete
POWER COMMAND
for Yourself
More power . . . more positive command over tractor and implement operation are yours in this new beautiful Ferguson "40". It's longer, heavier, can be used for mid-mount cultivation. Power steering and power wheels, too, if you wish. For your best tractor buy, see this Ferguson "40" and the three high-clearance models with convertible front-wheel assemblies. Remember: every Ferguson Tractor has the Ferguson System.
DEALER'S NAME
ADDRESS
PHONE

This is one of the latter day advertisements for the Ferguson 40. Trade that old tractor in – notice it's not an old Ferguson! March 1957.

March 2, 1957 THE FARMER

Down payment . . . one old headache

This farmer just made a real buy. He used his old tractor as down payment on a beautiful new Ferguson 40.

It's a trade where the farmer can't lose. Because now—at low cost—he has a tractor that's powered to save job time. The Ferguson plows with 3 bottoms in most soils, using much less gasoline. The Ferguson hitch saves time in the yard and in the field. So does maneuverability. And advanced hydraulic control speeds up closer work—whether it's cultivating, grading or contour plowing.

"It's got the Ferguson System," is the way most owners explain how they can do so much with this tractor.

No fiddling with complicated adjustments. No wasted fuel. You get full use of power—automatically.

These are facts which can be pretty pleasant when you total up the result of a year's work. There's no need to let an older tractor hold back your farm profit. Talk to your Ferguson dealer. Talk performance. Talk trade. Perhaps your tractor, too, will be the down payment. Make it a special point to see your dealer right away. See him first . . . or see him last. Either way you can't lose. *Ferguson, Racine, Wisconsin.*

SEE YOUR FERGUSON DEALER

see him FIRST . . . if you want the best deal but don't have time to shop around.

or see him LAST . . . if you still want to shop around and convince yourself.

EITHER WAY YOU CAN'T LOSE

Ferguson

POWERED TO SAVE YOUR JOB TIME